ONCE WE ALL HAD GILLS

ONCE WE ALL HAD GILLS

Growing Up Evolutionist in an Evolving World

RUDOLF A. RAFF

INDIANA UNIVERSITY PRESS *Bloomington & Indianapolis*

This book is a publication of

INDIANA UNIVERSITY PRESS
601 North Morton Street
Bloomington, Indiana 47404-3797 USA

iupress.indiana.edu

Telephone orders 800-842-6796
Fax orders 812-855-7931

⊜ The paper used in this publication meets the minimum requirements of the American National Standard for Information Sciences – Permanence of Paper for Printed Library Materials, ANSI Z39.48–1992.

Manufactured in the United States of America

Library of Congress Cataloging-in-Publication Data

Raff, Rudolf A.
Once we all had gills : growing up evolutionist in an evolving world / Rudolf A. Raff.
p. cm.
Includes bibliographical references and index.
ISBN 978-0-253-00235-8 (cloth : alk. paper) – ISBN 978-0-253-00717-9 (ebook) 1. Raff, Rudolf A. 2. Biologists – Biography. 3. Evolution (Biology) I. Title.
QH31.R128A3 2012
570.92 – dc23
[B] 2012005742

1 2 3 4 5 17 16 15 14 13 12

TO BETH, AND TO ALL MY FAMILY, NEAR AND FAR

Contents

Preface

MOST AMERICANS YOU MIGHT MEET ON THE STREET COULD name at least one living athlete, musician, celebrity, and politician, but far fewer could name any living scientist – or tell you what scientists do. A British poll of teens found that none could name a single living scientist. A poll of Americans found that fewer than 20 percent of respondents could name one. Sadly, despite living in an age defined by science and what it has produced, most Americans are content to enjoy the benefits without much intellectual engagement. Science is deeply integrated into the pillars of our culture, and on the flip side it is also part of political disputes over what is taught in schools and how public decisions are reached about energy policy, conservation, population size, contraception, vaccination, global warming, stem cell research, nuclear weapons, and many other issues that influence our lives and futures. When the public is befuddled, these issues will be decided in ill-informed ways based on religious, economic, and political biases that ignore the realities of the natural world.

What do we do about global warming when the public understands the science only vaguely at best and can't tell the real science from the rosy predictions of cranks or the self-interested lobbying of paid denialists? Worse, many of our fellow citizens don't think it matters anyway. Part of the problem lies with us, the scientists. We are excited about science and we enjoy our work. We discuss it exuberantly with our colleagues, but the voices of scientists talking among themselves don't carry very far. We should reach out and make the effort to talk to more than just our research colleagues, to tell a personal story of where we came from, why we are compelled by science, and what our own work is about. Some scientists have done that with real verve. There are several "autobiographical science" books that have inspired me over the years, including Charles Darwin's *Voyage of the Beagle*, L. S. B. Leakey's *White African*, Marty Crump's *Searching for the Golden Frog*, Margaret Lowman's *Life in the Treetops*, and Geerat Vermeij's *Privileged Hands*. Some scientists see this kind of writing as mere "popularizing," but popularizing broadens public understanding and certainly doesn't conflict with doing good science. This book became my own effort to speak up for science, to tell how I became an evolutionary biologist and explain, in part, the vibrant and living science of evolution.

The day I began writing this book, I really hadn't fully thought out that goal. I imagined that I'd like to write a short history of my life for family, siblings, cousins, children, and our new grandson, Daniel. I started out wanting to leave a clearer record of memory for my own children than I had gotten from my parents and grandparents. They left behind a sparse and disconnected record, photographs and snippets of conversations that stand like isolated rock pinnacles in the desert. They are dead now, and I can't ask them all the questions I should have asked years ago about their lives and origins. I've had to reconstruct and try to understand some of their histories as I went along. The scope and scale of the project simply grew as I got engaged. The writing became more complex, and I began to see it as something more than just a personal story for my family; rather, it transformed into a wider statement about how and why I became a scientist, the joys of doing science, and the importance of science for "real life."

Evolution provided me with the way to tell the story. Evolution is what I study, and it touches all that is alive. It is the science of origins and transformations in the history of life. This book is in part a story of my own evolution, of how I evolved into a naturalist, a scientist, and, finally, an evolutionary biologist. But more, it is a story of evolution writ large: its history, how it's studied, and what it means. My own life is a kind of thread through that story. The evolution of life has produced us, and all living things. It is hard to imagine a more compelling subject, yet evolution is also endlessly contentious. In America evolution has become an enemy useful to help organize a religious culture war being waged against rational thought, and even against the concept of a secular country tolerant of all religious views – the foundational notion of the separation of church and state. That conflict is also a part of the history of evolution. Its outcome may be decisive in whether our species will successfully cope on a changing planet. Evolution can't be ignored.

The bibliography at the end of the book is a selected one, meant to help any of my readers who might want to look further. I didn't want to create a distracting academic list of citations that would interrupt the flow of the narrative, so I've limited the bibliography to include two kinds of references. The first are citations of publications where really needed, for example, the source of quotations or crucial factual information used in the text. The second are books or articles that provide important and interesting narratives that make up part of the underpinning of the book.

Acknowledgments

NO WRITING CAN BE DONE IN ISOLATION FROM THE INFLUENCES and assistance of others. The greatest help I've had is from my wife and scientific partner Beth Raff. I am deeply grateful for her love and encouragement, for her willingness to read endless manuscript versions, for the clarity of her thinking, and for her wielding of a merciless red pencil. Her comments exposed embarrassingly bad organization and muzzy thinking wherever they hid among the grass of unobjectionable words in the acreage of manuscript pages. My sister, Mimi Jakoi, and my cousins Monique Dufresne and Michelle Ricard shared their memories of our shared childhood and helped me find family histories. I thank Ed Fraser and George Glauber for providing me with information that helped me understand my father's emigration to Canada from Austria in 1938.

Many people appear in this book, including family, friends, students, teachers, and colleagues. I remember them all for enriching my life and intellectual growth, and the science I write about. I owe a great deal to my academic home, Indiana University, which has made it possible for me as a faculty member in biology to work in a rich and intellectually exciting environment among admired colleagues. I also have to acknowledge a parallel universe at the University of Sydney. The School of Biological Sciences at Sydney has generously allowed Beth and me, and our students, to do field work and research in their facilities each year since 1986. The generosity of the school, the heads of school who served during the time we worked in Sydney, and so many colleagues in and out of biology in Sydney has given us the ability to conduct a research program dependant on evolution in the lost continent. This story could not be told without all that our Australian friends and collaborators have done with such grace and zest. I'm especially grateful to Don and Jo Anderson, who made it possible for me to start my research in Sydney. Heather Sowden, Craig Sowden, Hamlet Giragossyan, Margaret Gilchrist, Les Edwards, Mark Ahern, Michael Joseph, Basil Panayotakos, Malcolm Ricketts, and many other members of the Sydney faculty and staff have selflessly helped us make our field and lab work possible. Maria Byrne has been a priceless long-term collaborator. I thank Robyn Stuchbury and Noel Tait for all our times spent together in Australia, for teaching us about Australian biology and providing a superb color photograph of a living peripatus. The Women's College at the University of Sydney has provided a welcoming home away from home to Beth and me and our students for over two decades of our research. I am also grateful to Haris Lessios of the Smithsonian Tropical Research Institute for hosting us on a crucial research trip in Panama.

As always, as scientists we owe an enormous debt to our graduate students, postdoctoral fellows, and technicians, who throughout our careers have made meaningful research possible. I acknowledge that debt. Too many people have been part of the effort to list them all here. Some are cited specifically in the book and in the bibliography. I want to thank Mary Andrews, whose technical abilities made much of our Australian work possible, and Pat Anderson for her outstanding work as editorial assistant for *Evolution & Development* and for her help with the manuscript of this book. Cited or not, I prize you all.

I want to acknowledge Editorial Director Robert Sloan and the staff of Indiana University Press, who have done so much to make the book a physical reality. I thank Angela Burton for guiding me through the process, and Marilyn Augst for the creation of the index. I am indebted to the external reviewers of the book manuscript, Brian Hall and Billie Swalla, for their comments and advice. Finally, I thank Tom Jorstad, Smithsonian Institution, for arranging the use of a photograph of a 500-million-year-old precursor of living peripatus in the fossil record. I am grateful to colleagues Phil Donoghue, Shuhai Xiao, Deiter Waloszek, and Ron Blakey for supplying images from their scientific works, and to paleo artist Raúl Martin, who made available his painting of *Acanthastega* for use on the cover. I thank Jim Gehling and the South Australian Museum for showing us Ediacaran life and preservation through the museum's spectacular fossils.

Beth has reminded me to say that any surviving errors are mine alone.

ONCE WE ALL HAD GILLS

PART ONE

Becoming a Naturalist

WE DON'T OFTEN TAKE THE TIME TO TELL PEOPLE WHY WE are scientists and how we developed intellectually. I think that's a mistake in a time when science influences society heavily but few people know a scientist or what he or she does. No one speaks for all scientists, but I can tell you my perspective. The first part of this memoir is my attempt to unravel my past and decipher how I came to be a passionate naturalist as a child and eventually into a fledging scientist trying to decide what kind of science I wanted to spend my life on.

Very few things happen at the right time, and the rest do not happen at all. The conscientious historian will correct these defects.

Herodotus

ONE

Space-Time

A BRIEF HISTORY OF A LONG TIME

I am enthralled by time. As long as I can recall I've wanted to know how the familiar world we take for granted came about. This has been a lifelong fascination because the past is truly not just another country but a chain of linked and ever stranger other worlds. Our evolutionary origins lie in these former worlds, which grow not only more alien but also fainter and more elusive as we look ever deeper. The passage of years and the eclipse of memory also obscure our personal origins. Like detectives, we have to tease out our pasts from imperfect and concealed evidence. On the greatest earthly scale, the geological record of the planet and the record of the evolution of life upon it have been likened to a book left to us with most of its pages torn out. On a personal level we suffer from lost family records, deceased witnesses, and the erroneous illusion that our own so-certain memories are accurate. Our efforts to answer questions on these vastly different scales will succeed with some, but others will remain elusive, and new questions will arise like dimly seen specters, shyly but persistently standing at the edge of our vision.

To start somewhere, I'll begin egocentrically with the place I was born, the Quebec city of Shawinigan. Throughout my life this obscure town has remained in my imagination not only as my birthplace but also as a symbolic dividing point that marks the end of the road with the unknown North of bleak arctic Canada stretching into the time and distance beyond. Shawinigan has no long pioneer history, and its founders were not noble frontier settlers living in rough log cabins, bravely hunkering down in isolation to weather the long winters. This was a town founded in 1901 by a hydroelectric power company that exploited the roiling water of Shawinigan Falls to turn its generators. The city had grown to twenty thousand people by 1941, the year I was born. Even now, only the road to La Tuque runs north out of town, and I have only seen the country lying far north of Shawinigan from thousands of feet in the air, gazing out from flights from Iceland or Europe that cross over the glaciers and icebergs of Greenland and over northern Canada.

Quebec is a beautiful but hard country, hard in its climate, hard in its bones, its people enduring. Its history starts with the enterprising explorer Jacques Cartier staking a claim to the place by planting a cross on the Gaspé Peninsula in 1534. French pioneers colonized Quebec for its value in trading steel tools, firearms, and dyed cloth for beaver pelts with its original Algonquin and Huron residents. Quebec City was founded in 1608, but the promising French empire in Canada ended in disappointment in the next century at the end of the Seven Years' War. French Canada was ceded to Britain in the Treaty of Paris, signed in 1763. The British government attempted to assimilate the Québécois and blocked Catholics from holding office. Not surprisingly, the policy caused alienation, and it was abandoned with the Quebec Act of 1774, which restored religious rights and French civil law to French Canadians. When American revolutionary armies invaded Quebec in 1775, French Canadians did not support them. Québécois stubbornly remained French speaking and Catholic, with religion a force of conservativeness as well as of unity and cultural salvation. In the mid-twentieth century, Quebec began to liberalize, and to play a lively role in Canadian politics with struggles over its status, control of its economy, and even whether to separate from Canada – from the Plains of Abraham to Quebec libre. It's not clear where that's going, but at least road signs across Canada are now bilingual, which I guess looks quaint to Americans. Several Canadian prime ministers have been from Quebec, even one, Jean Chrétien, from Shawinigan.

The map of Quebec covers an area over twice that of Texas. It is endowed with a geological setting that speaks the origin of the world. Its surface makes up part of the Canadian Shield, a land of ancient crust, bent, heated, compressed, turned into tortured stone stretching in bare places as great tapestries of light- and dark-banded schist. We see a surface produced by the scouring of mile-thick Ice Age glaciers only a few thousand years ago. They left behind smoothly polished hard rock pavements, and the rocks record many insults over the ages. Large meteor craters remain as scars indicating where millions of years ago mountain-sized asteroids plunged into the already ancient Canadian crust. From the right-hand side of the plane on flights coming south from Greenland, one can see three large meteor craters in northern Quebec, the paired Clearwater Lakes and Lake Manicouagan, which are sometimes visible below. Manicouagan crater is one hundred kilometers in diameter and visible from space as a prominent ring of water. The crater was blasted out by an asteroid or comet 212 million years ago. Farther south are the old and rounded Laurentian Mountains, composed of twisted billion-year-old metamorphic rocks. The vast scattering of blue-water lakes across Quebec is the glittering gift of recently melted ice sheets. Laurentide Quebec is largely greened by expanses of forest, the maple, spruce, fir, and decorative white birch woods in which that I spent a part of my childhood.

Admiration for one's own birthplace is a common form of narcissism, part of our highly evolved human self-centeredness. Still, the Canadian Shield, a part of a primeval continent that geologists call Laurentia, has cachet as one of the first regions of crust to have formed on the cooling early Earth. Rocks on the Canadian Shield are among the oldest found on Earth (4 billion years – not bad considering that the age of the Earth has been estimated at 4.55 billion years from dates obtained for meteorites). To put these numbers in perspective (as if we, who suffer pangs of eternity when we spend three hours in the molded plastic seats of an airport terminal waiting for a delayed flight, could really have any sense of such spans of time), astronomers have determined that the universe came into being about 13.7 billion years ago in the "Big Bang." The oldest crust in Laurentia is just shy of a third the age of the universe. The events leading to that first bit of Canadian crust reach back to the earliest giant stars and their violent ends as supernovas, explosions that created the heavy elements of the universe.

The dust and gas surrounding the early Sun came from the violent deaths of the first two extravagant generations of stars of the young universe and supplied the materials necessary to form the solar system. The primordial stuff included the rocky elements of our planet – silicon, calcium, aluminum, potassium, sodium, and iron. It also included the oxygen that combined with hydrogen to make the water of the oceans, and with silicon to make the hard clear crystals of quartz. The atoms of carbon and other elements that formed the molecules of life were themselves born in supernovas. Planets formed through the accretion of the disc of dust circling the Sun. The Moon arose through a violent collision of the proto-Earth with a Mars-sized neighbor. Then following a final drumbeat, a spasm of asteroid collisions, our planet's surface and oceans appeared and geology began to keep its long record.

Laurentia was one of the first continental plates to form as things calmed down and a stable planetary surface emerged, but none of this is to suggest Laurentia has been unchanging for the enormous time since the days of its birth. Its history is more a story of endurance in the face of continuing inexorable and sometimes violent change on a geologically active planet. The old crustal fragment is a remnant of an Earth before life, of a world with no oxygen in its air, a world with a recently formed moon that looms close and large in a smoggy sky and causes enormous tides. It is a world that is just beginning the geological cooking that will fractionate the lighter density rocks that produce the continents from the heavier basalt of the ocean floors. It is a world where about the same time as the birth of the Canadian Shield, life appears by 3.4 billion years ago and begins its billions of years of biological modification of the atmosphere and oceans.

For an agonizingly long time, all life on Earth was made up of single-celled microbes that thrived without oxygen. Then, 2.5 billion years ago, single-celled organisms that harvested energy from sunlight started transforming the atmosphere into one that for the first time contained oxygen. At first, oxygen was a pollutant, a waste product of photosynthesis spewed out by green bacteria called cyanobacteria. Oxygen was a lethal poison to the other life forms, the anaerobes that existed up to that time. The emergence of an atmosphere with oxygen would in time change the composition of the air and cause a revolution in the history of life. It eventually made possible the emergence of all the life we see on the Earth's surface, because it provided ample oxygen for highly efficient aerobic metabolism and allowed the production of ozone (O_3), which blocked deadly ultraviolet light from blasting any organism that ventured to show itself on the surface. Yet an atmosphere flooded with oxygen at levels similar to those of our present atmosphere did not finally appear until about 700 million years ago. Animals then quickly evolved in the newly oxygen-rich seas. And yet, microbes that can't stand oxygen did not become extinct. They are still with us, and doing just fine as microbes always seem to do, but they have retreated to inhospitable environments such as the airless black mud of bogs, the inside of our guts, and the vastness of the deeply buried rocks that lie under the continents and beneath the ocean floor. Nonetheless, these old anaerobic creatures may make up most of the mass of life on Earth. Bacteria have been recovered from South African gold mines at depths of almost three miles. A vast hidden ecosystem of bacteria that have their own energy sources and no contact with the world of the surface lives in the abundant rock of the crustal underworld.

Laurentia's geology also records the growth of the continent through long-ago additions of pieces of crust that collided with the continental core. The collisions threw up great ice-covered peaks formed out of rocks rumpled like layers of putty by the inexorable force. In the vastness of time, these mountains have been weathered away by ice and water and wind, grain by grain into gravel and sand and mud. The polishing of recent Ice Age glaciers, which ground and scoured the continent, reveal the remnants of the compressed and folded roots of the ancient mountains. These rocky hieroglyphs are the only witness we have to the presence of those billion-year-old peaks. The continental glaciers melted away only about ten thousand years ago. With the retreat of the ice, the trumpeting of the last mammoths faded away from the warming Quebec tundra.

A history of magnetism frozen in once-molten rocks as they solidified shows something even more remarkable. Laurentia itself has moved, and through three billion years it has danced slowly over the Earth's surface along with the other continents, now colliding to make a larger continent, now breaking up into smaller continents. The continents drifted in position over the globe through the inexorable power of plate tectonics driven by the internal heat of the Earth. This process continues. Europe is presently moving away from North America as new crust erupting at the Mid-Atlantic Ridge expands the sea floor and carries the two continents apart – at more or less the rate at which fingernails grow. About 200 million years ago, when the Atlantic Ocean began to form, Europe and North America were fused and still part of a huge northern supercontinent. Going further back in time to the age when animal life was diversifying, about 500 million years ago, what is now Quebec was located just south of the equator.

A LONG HISTORY OF A BRIEF TIME

As for my more recent origins, I know the place and date of my birth, but not from a birth certificate. That never existed. My start in life took place in the gloomy autumn of 1941, shortly before Pearl Harbor, and is recorded officially on a handwritten baptismal certificate in the name of Rudolf Albert Joseph Raff. A few days after the event, this document was filled out in hand by my mother's parish priest and entered the church baptismal record. What a medieval kind of approach to keeping birth data in a modern country. How inconvenient it would be for the unbaptized. It seems so early for an ambiguity to have appeared in my own documentary fossil record, a tenuous connection to my actual birth. Rudolf came from my father's name, Albert for my grandfather Dufresne, and Joseph was my "saint's name." The saint's name was later mercifully dropped as an excessive frill on the U.S. side of the border upon my naturalization.

My father, Rudolf August Victor Raff, was born in 1908 in Mödling, Austria, a modest village under a scenic mountain twenty kilometers southwest of Vienna. Gerd Müller, a professor of zoology at the University of Vienna and a generous host, took my wife Beth and me there while we were visiting the Conrad Lorenz Institute as his guests in 2006. It was a visit of curiosity and trepidation. Mödling has narrow streets and one of those strange and elaborate early-eighteenth-century "plague statues" that commemorate the last visit of the bubonic plague in 1713. These statues have an intricate symbolism, concoctions of grimacing skulls and dying victims topped by a cheerful angel. The plague swept its scythe through the people of Mödling just a generation after the same treatment by the Ottoman invaders who laid siege to Vienna in 1683. The old townspeople had managed to express hope in that angel. In the living town that hope has evolved into an excellent coffee and pastry shop with tables on the town square. We easily found the building at 20 Elizabeth Strasse, where, as my father's birth certificate records, his parents lived when he was born. It felt strange to be standing there at this intersection of past and present just two years shy of a century after my father's parents first brought him home. It is still an apartment building and looks much like a family picture from the early twentieth century.

My father took his Ph.D. in chemistry at the University of Vienna in 1932. His intellectual roots lay in the rich history of science in Vienna, and his dissertation committee included the eminent paleontologist Othenio Abel. Abel was one of the founders of paleobiology, the study of how extinct creatures we know only as fossils once lived. Abel seems to be most remembered for proposing that fossil skulls of dwarf mammoths found on Mediterranean islands may have been the source of the Cyclops legend of the ancient world. Perhaps the downside of fame is to be remembered for the trivial. The family liking for paleontology apparently ran at least one generation deeper. Although my grandfather was a lawyer, as a student he had some interest in science. I have a modest geological hammer that my father told me had been given to his father when he was a student at Vienna by the geologist Eduard Seuss. This was the same Professor Seuss who became famous for proposing the existence of an ancient southern supercontinent he called Gondwanaland after the Gond tribe of India. Gondwanaland (now called Gondwana) is an important component of understanding continental drift. That little hammer has drifted across time, an ostensible relic of ancient rocks once lovingly pounded by the great geologist.

My grandfather's name was Rudolf Ignaz Raff. My grandmother was Emma Steidler Raff. Her father was apparently a physician. I have no history of my father's Austrian family before them, and few stories even of my grandparents' lives. Judging by photographs of my father (an only child) with his parents, they were somewhat or, I suspect, quite reclusive people. My father had a cousin, but in the late 1930s the family was bitterly split by his cousin's support of Hitler. Contact never was reestablished after the war. When Beth and I were in Vienna in 2006, I looked in the city telephone book for any Raffs. I expected to find dozens, but there were only four listed; it was seventy years later and I couldn't bring myself to try dialing any of them. I could only imagine that awkward English-German phone conversation. As for more remotely possible ancestors, my father alluded to the Swiss composer of syrupy music Joachim Raff as a possible relative. Joachim was born in 1822. However, he makes a poor ancestor as his only child, a daughter, left no descendants – so surely this story is just one of those family legends and elusive false clues.

As part of the former Austrian government before the German takeover, my grandfather was in disfavor with the Nazi regime and unable to work after the merger of Austria into the Reich. He died of malnutrition in 1943. My grandmother sometimes had to cross into Hungary to buy black market food from farmers willing to sell produce. Ironically, she survived the war because German army officers occupied part of her house and kept her fed. Russian officers in 1945 and for a brief period of occupation following the war took over her house and did the same. I knew my grandmother only for a short interval, when she came to live with us in Cornwall, Ontario, after the war. It didn't work out. She spoke no English and was unhappy to the point of depression. I can remember sitting in her room watching her endlessly brush her pale long hair. After a few months she returned home and lived in familiar, comfortable Austria the rest of her life. She had occasional visits from my father, and once each from my mother and sister. In one visit to her my father took my sister, then in her late teens, with him. They went to a convent where his father was buried and he asked my sister to wait for him at the *Kinderdorf* (children's home) while he went to the gravesite alone. None of us now know where that is. My grandmother died in 1963. Unfortunately, I was not able to visit Austria while she was alive, but I inherited the pendulum wall clock she and my grandfather bought in 1916. It outlived all but faint memories and still runs.

In the spring of 1938 my father decided to escape the worsening social and political situation in Vienna that followed the Anschluss of Austria with Germany and the sweeping Nazification he despised. He told me that he would walk to work on streets that would not force him to pass a portrait of Hitler on a building, because anyone walking by was required to salute Hitler's picture. William Shirer mentions in his *Berlin Diary* that failure to salute Nazi street icons could get even foreigners beaten up. My father left Vienna by dint of taking a temporary position working in a Canadian chemical company. His departure followed that of his mentor, Professor Herman Mark, who was famous as the founder of polymer chemistry and was the creator of the polymer chemistry program at the University of Vienna. I asked a friend of Jewish Austrian descent, who fled Austria as a child with his family, about his experiences after the Nazi takeover. He told me harrowing stories of his immediate family. Some escaped; some did not. Austrian Jews regarded themselves as Austrians first, but that made no difference. Those who survived did so because they saw their fate soon enough to leave before being rounded up, because of sheer luck, or, in my friend's father's case, because of unanticipated help after he was interned. At that time, the rules allowed internees a twenty-four-hour leave for a family emergency. His wife had written that one of their children was sick, and he took the letter to the camp commandant, a Luftwaffe officer and fellow Great War veteran. It went like this. The commandant asked, "As one officer to another, is this true?" My friend's father replied, "I can't say for sure." "Alright," said the commandant, "I'll grant you the leave – I wish I were going with you." The family went as refugees to Belgium and piecemeal got visas to the United States.

By 1938 Professor Mark clearly saw how events would unfold and decided to leave Austria. Because he was Jewish, Mark had been arrested immediately by the Gestapo after the Anschluss. He was released from their care, but his passport was confiscated. That he got back with a hefty bribe and was able somehow to get a visa enabling him to go to Canada via Switzerland. As a chemist Mark knew how to buy platinum wire – tens of thousands of dollars worth. He secretly made the wire into coat hangers, for which his wife knit covers so the sheen would be a little less obvious. He fled Austria in a creative and dramatic fashion by loading his family, with clothes on their precious hangers, into the family car, tying a swastika flag onto its radiator, strapping the family's skis to the top, and driving to Switzerland on a "ski holiday." I imagine the Austrian border guards took him to be a Nazi big shot, saluted smartly, and lifted the gate into Switzerland. I heard some of this account from my mother, who was fond of the tale about the platinum coat hangers for its sheer impudence. I learned most of the story only recently from reading accounts of Mark's life. Once again, family stories were either fragmentary or woefully badly remembered by me. From Switzerland he went to Canada and then in 1940 moved to New York.

One wonders, how do you know when it's time to leave? Was my father moved to emigrate when he did because of the way Mark was abused by the Nazis? It must have been shocking to see his teacher, an eminent scientist and World War I Austrian war hero treated in this absurd and brutal fashion. Mark and my father went to Canada within a short time of each other, and I suspect that he helped my father make the connections that led to his offer of employment in Canada. They co-authored a book, published in 1941 and titled *High Polymeric Reactions*. The title page of their book acknowledges its translators, which indicates that it was drafted in German and translated into English. Mark and my father must have written the manuscript in 1939–1940 while they were both in Canada. It is a comment on the remarkable degree of Canadian confidence and good sense that these two displaced Austrian scientists were sheltered and allowed to work freely in the midst of wartime.

Why Quebec? Simply, my father spoke French well, but not English. His classical education included several years of Latin and Greek. Unfortunately, these languages were not all that handy in twentieth-century North America. His transatlantic trip was via a British passenger ship, which caused him what seem like preposterous tribulations. The combination of his limited English vocabulary and his reluctance to try out anything unknown on the menu condemned him to eat eggs on toast for breakfast morning after morning. He never ate fried eggs again, and he made sure of becoming highly proficient in English once he lived in Canada. Two events in this migration created the contingencies necessary for my eventual existence. First, the Austrian official who granted my father an exit visa evidently had not paid attention to arithmetic in school and made a mistake in the return date, with the outcome that my father was allowed a two-year period in Canada rather than the officially approved single year. The propitious second event was that his gangplank dropped him off at the Shawinigan Chemical Company research lab. There he met my mother, who was from Shawinigan and worked in the company office. They were married in November 1939. Perhaps because he married a Canadian, and because administrators and colleagues from Shawinigan Chemical wrote strong letters of support that said he was anti-Nazi and pro-British, my father (by then officially an enemy alien) was not interned and continued to be employed as a scientist in Canada. The rest for me was biology – and baptismal record keeping.

There remained something of a puzzle about my father's contact with his parents once World War II had begun. Canada and Austria were enemy combatants. I knew that somehow my parents had kept in touch with my grandmother and had sent her CARE packages. The answer came literally out of the blue (actually, via email) from a stranger. In 2009 I got a message from Ed Fraser, a postal enthusiast and historian living on Long Island who had gotten hold of one of my grandmother's envelopes addressed to Post Office Box 252 in New York in the summer of 1941. The enclosed letter had not slipped by German censors and had been either returned to my grandmother or destroyed. The empty envelope survived and ended up in a postage stamp auction many years later. Out of many lost envelopes, this one fell into Fraser's hands. He told me that he thought the name Rudolf Raff was unusual enough to try a search on Google, and I turned up. He sent me a photograph of the envelope, which has the name Emma Raff on the reverse side in her characteristic handwriting.

Fraser wrote up the story of Box 252 in a journal called the *Posthorn*. With the start of the war, Canada became a combatant, but the United States remained neutral until December 1941. Thus the United States was for a while a neutral transmission point for mail between families in the hostile countries. The U.S. Post Office had arranged with Thomas Cook and Sons in Toronto, where my parents then lived, to set up a neutral country message drop in New York City. The cost for sending mail this way was high for 1941 at fifty cents per letter. Because of wartime conditions and censorship, the turnaround time for correspondence could take up to a year. Points of censorship for a letter coming to Canada lay with the Germans, the British in Bermuda, where messages were passed to the United States, and finally in Canada. My grandmother likely had to stop sending letters via Box 252 in the summer of 1941 because a warning stamp from the German censor on her envelope says she would be reported if she again wrote to an enemy country via a neutral post drop. Messages were supposed to be sent from Red Cross to Red Cross. Text was limited to twenty-five words and to family matters. Only a handful of Box 252 envelopes survive.

They are not totally extinct. In some of us they live on, a little bit.

Svante Pääbo

TWO

Layers of the Past

HISTORIES

My mother, Therese Dufresne, was the daughter of a well-liked local physician, Albert Dufresne, who practiced from 1930 onward in Shawinigan and the surrounding countryside. His house calls could mean anything, including grueling trips into the backcountry by horse-drawn sled or canoe. By the time he retired, my grandfather had delivered or treated most of the living citizens of the town. He once estimated that he had delivered eight thousand Shawinigan babies. In 1966 he was made a Commandeur de l'Ordre de Saint-Gregoire-le-Grand, a papal award for his charitable acts to his many patients unable to pay in hard times. A street in Shawinigan now bears his name.

Going to visit Shawinigan during summer vacations was the highlight of my early life. Shamefully, it was not because I liked spending an entire vacation in my grandparents' rather formal house. I was too energetic for that. What I cherished most was any time I could spend out in the woods at a lake. My Uncle Gérard Dufresne's family had a remote cottage on Lac des Îlles, where on one visit I was impressed to see the hole where an enterprising bear had clawed its way through a soil-filled double-log wall into the icehouse. What a frisson to realize that wooden doors would be as paper to hungry bears (not that they bothered cottages with people around). Most of my cottage experience though was at Lac Souris (Mouse Lake – had they run out of better names?). Here the vast Quebec forest lapped the edge of civilization. On the far side of the lake, inaccessible from the end of the rutted lake road, my uncle Guy Ricard (the husband of my mother's sister Margot) and my grandfather had built a summer cottage. To get to the cottage from the road head, we would uncover my grandfather's old motorboat, drag it over the wet sand into the shallows, load up supplies and gas, and push off with battered oars to get into water deep enough to lower the outboard. Then, with some boat rocking, repeated pulls of the starter cable finally got the balky engine going. We'd head off at two miles per hour in a cloud of fragrant blue smoke. If there were just the two of us, I'd be allowed to run the engine and steer with my grandfather's nervous guidance. Once steady, I could throttle up enough to leave a discernable wake across the usually glassy surface. I have a photograph of one of those days – me a skinny ten year old wearing an oversized old raincoat of my grandfather's belted around my waist, he with his inevitable cigar in his mouth.

The center of Lac Souris had a darkness and chill born of what we children thought of as an immeasurable depth. With the green conifer wilderness, the lake beckoned with promised mysteries and adventure. The cottage sat just a short distance from a yellow sand beach with an ebb and flow of glittering golden mica flakes that played in the ripples. My daily companion in splashing around was my cousin Pierre Ricard. Every day was an unlimited and unsupervised opportunity for swimming in that frigid lake – an option we exercised fully. Of course, we didn't know any better than to shed body heat into the cold water till we emerged shivering, numb, and clammy. We'd quickly change in the attic and warm up by the stove. There was no electricity. The only drinking water came from a rusty hand pump that had to be primed with water from a pail each time it was used. The well water it brought up had a powerful and nasty mineral taste that we were told was good for us. And there was plenty of practical entomology – black flies in June, horseflies in August, and humming clouds of mozzies to fill in for the rest of summer.

Pierre's sister Michelle reminds me that Pierre and I made the younger sibs and cousins play extreme hide and seek games in the woods, and in one such game we ambushed them with an avalanche of boulders. My brother tells me it was a giant log we cast down – something several feet long that crashed through the foliage. I remember the avalanche as a little heap of gravel and small cobbles pushed off a six-foot bank at the edge of a wide logging trail, all to the accompaniment of what we thought were convincing bear noises. That kind of play between ages seems a universal. Beth, also an oldest child, has told me that in her farm summers she and her older cousins kept the younger kids in her family from wanting to share in riding horseback by making their mounts Gypsy, Topper, and Molly ostentatiously buck by pulling in their reins while kicking their flanks. These contradictory signals made the horses shake their heads and prance, altering gentle farm horses into fictive dangerous snorting broncos for the younger sibs' edification. Mission accomplished.

Every so often other relatives would show up to spend a brief day at the lake. Mostly these occasional woodsmen took in the sun, lounging in canvas deck chairs on the cabin porch. Then they had lunch, and soon headed back to the comforts of civilization and indoor plumbing minus black widows. I'm grateful to one of these visitors though, my amiable, voluble and vigorous black sheep uncle Eugène Dufresne (my grandfather's brother), whose tales of adventure included accounts of time-intensive schemes for smuggling miniscule amounts of cigarettes and whiskey across the U.S. border for the thrill of outwitting customs. Well into his seventies he boasted about his new "lady friend," whom we never met. One cold cloudy day at the sad chilly end of the Lac Souris season, as the cottage was being readied for winter closing, he invited me out in the motorboat (five horsepower and slow) to shoot fish with a shotgun he had brought along. I'm not sure what possessed him to suggest this particular sport, but I was bored and game for a try. I sat in the stern and ran the outboard at a painfully slow, nearly idle, crawl over the slate gray lake. Eugène crouched in the bow like a trench coated ancient mariner, a make believe whaling ship sailor grasping his small-scale harpoon gun. A large lake trout jumped. He fired the shotgun. A boom echoed over the lake and the pellets hit empty water where only the latent memory of fish lingered. We puttered around circling in the drizzle a while longer and went back in to the dry cottage. I'm sure Eugène would have tried dynamite if it had been available.

My mother's family reaches far back into Quebec history. The earliest ancestor she claimed, as I recall the tale, was Madelain de Verchères (1678–1747), a colonial heroine. As a teenager Madelain took charge of a weeklong defense of the wilderness family fort against an Iroquois raid.

According to the Marianopolis College Quebec history site, where I read up on Madelain, "In her later years she was chiefly distinguished on account of the large number of law-suits in which she engaged." Tedious long cold winters, no TV, no cell phone – who can blame her for suing the neighbors for diversion? They might have welcomed a little diversion themselves.

On my grandmother's side, our immediate ancestors were named Milette, who, in my memory of the family story, derived from a minor French nobleman who opportunistically changed sides once too often in the period following the French Revolution. It's hard to be critical – avoiding the guillotine is an effective evolutionary strategy. After the restoration of the Bourbon king Louis XVIII, he found another longevity promoting move in departing the Old World for Canada. Unfortunately, the Catholic Church in Quebec ended up in possession of his considerable property holdings after his death, apparently granting him a passport or fire escape to heaven in return for his deathbed generosity. He seems to have been ever the survivor. Although the family was staunchly Catholic, this holy shake down has always remained a sore point. I have no evidence that it didn't happen, or that it did, although my mother said she remembered once visiting his resting place.

I have to say here that much of my account of these ancestors is based on my faint memories of bits of oral history that I heard as a not overly attentive child. I'm sure what I remember is incomplete and even garbled. Now, when I would like more details and to have questions answered, the people who could so easily fill me in are dead. I suspect a lot of family history is lost this way or enters a realm of family legend with some elusive and distorted core of truth. I do have some genealogy data on my maternal grandfather's side of the family acquired by my uncle Gérard. The Quebec Dufresne lineage begins with the birth in 1700 of Philippe Dufresne in Normandy, France, and continues in true Old Testament begat style to the present.

My grandfather and grandmother, Laetitia Milette Dufresne, were Catholics. In tune with their time, they were given to a religious form that believed in dramatic miracles and in the absolute authority of the church. They prayed daily and supported a Catholic education. For a brief time when my mother was quite young, they sent her and her sister

Margot to a convent school in some neighboring village. There the sisters required that their charges undress in the dark and wear a shift when bathing. The Dufresnes brought their daughters home when they found that the girls were terrified and could not sleep at night. Throughout her long life my grandmother fretted about the Antichrist, whom she identified in successive world political leaders. She also worried about babies being snatched from their carriages by rogue eagles and by seeing us playing with Protestant children – English speakers and thus accessible as playmates. I never wondered about all that. She was wonderfully affectionate and kind, and the rest was the sort of minor foibles one accepted in adults. In fact, many years later, when I returned to visit them with Beth my new wife, a Protestant, my grandmother put aside her diffuse prejudices and was joyous and welcoming.

The old frame house of my childhood visits was large with a wooden veranda along the front and side of the house – an all-weather gerbil track for children. The basement was fascinating and forbidden, dirt floored and pervaded by a musty smell compounded of bare earth and inches of wood chips. There was a huge tree stump with an ax used for splitting firewood forebodingly imbedded in its center like King Arthur's sword. The smell of certain kinds of wood mulch still powerfully conjures up for me a strong memory of the look and feeling of that basement. A visit to my grandparents was always something of a visit to a past long vanished from where I lived. As late as the 1950s, ice blocks were still used in Shawinigan for home food refrigeration in thick-walled iceboxes. The iceman stopped by once a week with a horse drawn wagon containing foot-square blocks of ice cut from the St. Maurice River in winter and stored for summer sale. His coming was a spectacle heralded by sound of horses' hoofs and the creak of heavy wheels. When the iceman opened the door of the wagon it was to a cold cavern dripping with water. We stood in awe of the enormous iron tongs he used to shift the ice blocks from the gloom. In these innocent pre-giardia days, the iceman would give the gathered children chips of ice from amid the flakes of sawdust insulation to suck on. Imagine, horse drawn business wagons. Yet the iceman wasn't alone. Well into the summers of my high school years French fry vendors plied their trade from chip wagons standing behind drowsy horses.

The Dufresne kitchen was a large room that served as the family space, because my grandmother kept her dining room and parlor for formal occasions. The kitchen was a sunny voluminous place with wide windows framed by two cages of canaries, each standing on spread out newspapers. Her stove was a massive wood-burning iron device with concentric circular lids on its surface. These were opened by hand with a lifter that fitted a notch in the stove lids. Each morning the stove was lit using a scary accelerant (perhaps kerosene) before breakfast and allowed to smolder on all day to be used when needed. I was nervous of the morning firing up, but would never miss watching. On looking back, the greatest oddity about the layout of the house was that my grandparent's bedroom opened right off the kitchen. When my grandfather retired from his medical practice in the 1960s and sold the house, its new owners converted it into a funeral parlor. The veranda disappeared under a blank gray wall that made a once inviting exterior ugly and graceless.

One of my earliest photographs with my grandfather shows me at about three years old sitting on the bumper of a highly polished splendid black late 1930s vintage car with him holding my hand. That car was so special and off limits that I remember well the ceremony with which I was lifted up to sit on it to pose for that picture. I recall my grandfather as sentimental and a great storyteller. One such tale from the early days of his career was that he had been compelled by an emergency to go by canoe into the roadless lake country guided by a distraught woodsman to treat his sick wife. Fifty years later a man crossed the street to embrace him to tell him he was that woodsman and tearfully to thank him again. It would appear that this story had a better outcome for the patient than the one in which my grandfather was called out into the country to treat a man who had struck himself in the head with a double-bladed ax while splitting logs. His finest hour as a raconteur came the last time I saw him, in his eighties. He took me into his office and showed me his framed graduation picture with all the serious stiff-collared young members of his 1911 medical school class. My grandfather's satisfied chortle was "I'm the only one left." Alas, there seems to have been no tontine to enrich his survivorship. He gave me his rolltop desk, which had been in use to take blood pressure and write bills since 1914. We hauled it back home tied to the roof of the car with a liberal use of rope and many inelegant knots. For some reason, its classification caused a minor quandary for U.S.

customs at the border. We were pulled out of line and parked behind the customs shed. Finally the supervising agent decided the old rolltop was household goods from some era when I had lived with my grandparents. We agreed upon a fictional duration for the sake of filling the blanks on the paperwork.

If the written account is dim, I can trace something of another kind of family history from a powerful new method that needs no family records to reveal my mother's ancestry in Quebec and into the long past before. This is hidden evidence that lies in the DNA of our cells, a kind of ancestry we can trace dimly through our mitochondrial genotypes. Mine was determined by DNA sequence analysis to be haplogroup W by my former daughter-in-law, Jenny Raff, a molecular anthropologist who studies the relationships of long-buried people using the ancient DNA their bones still contain. Jenny was doing a routine inventory of the mitochondrial DNAs of everyone in the building to catalog possible contamination from living people working in the vicinity. Modern DNA is much less damaged than ancient human DNA, and thus contaminants are more easily amplified than ancient DNA, but individual sequences are distinguishable, so a contaminant is easily spotted. A little piece of history thus came to me as an incidental gift from a molecular housekeeping operation.

Haplogroup W is a particular mitochondrial genotype that extends back about twenty-five thousand years to the Caucasus. It is a rare genotype in northern Europe and reached into northern France, perhaps with the Vikings. Haplogroup W reappears in Quebec in the genetic histories of French settlers, mapping a long migration of people from the ancient Middle East into Europe thousands of years ago, and into colonial Canada. Mitochondria are passed on exclusively from mother to offspring. Inheritance of mitochondrial DNA thus follows the maternal line in each generation, and so a long line of maternal ancestors who transmitted this haplotype to me was revealed. Now that mitochondrial DNA sequence typing can be had relatively cheaply, internet clubs of people who share particular mitochondrial haplotypes have arisen – proving that there can be discussion groups about anything.

Just for completeness, I traced something of my father's long-ago genetic lineage by having my Y chromosome genotype determined through the National Geographic's Genographic Project, which is tracing the histories of movement of human genotypes across the world. You take the cheek swab; they do the rest. The Y chromosome is found only in men. My male chromosome lineage left Africa with the great migration of modern people about fifty thousand years ago. The carriers of this chromosome made their way into southern Europe from the Middle East in the Neolithic revolution of about ten thousand years ago. At a later unknown time, some remote male ancestor carried it into Austria.

MEMORIES

Then there is memory. For some reason, all my first direct recollections are of being outside, of going across a grassy park on a sunny day with my mother, of being blown along a beach by a high wind full of stinging sand and then sitting in the gloom with much of the family inside a collapsed tent while the rain drummed on it and the wind made the canvas billow and writhe and flap deafeningly, of walking around the wooded edge of a lake with my father and uncle Guy, and managing to rub insect repellent in my eyes despite being warned that it would sting. These are all scattered memories, but when my more continuous memories begin, they too are more often than not of being outdoors. The best of these memories are of walks with my father along the railroad tracks that followed the Saint Lawrence River near Cornwall, Ontario, where my parents had moved during the war.

The riverside was vibrantly alive to me and seemed like a jungle of spiky plants. I can remember walking there with my father. It was special to be out with him, because I would not see him during the day except on weekends. He had a wide background as a naturalist, although he was in no way motivated to be a collector – even of such things as a life list of birds. He simply told me about the insects, plants, and objects we saw as we tramped in the rampant weeds along the railroad line. All of these recollections of my early years are fragments, but it's not obvious why I should remember any particular scraps of memory and images. That's true of those walks as well. In one, we stood by a railroad flatcar sitting on the rails. My father's hand held chunks of pure white gypsum and black glossy coal that had fallen off of a train. He told me what they were and what they were used for. I already knew about coal as fuel. We heated our house with a coal-burning furnace, a cast-iron monster that required a noisy feeding by my father with a wide-bladed shovel each morning, so I'd seen piles of coal before. What impressed me that day by the flatcar was the fact that the stuff was formed long ago in some mysterious way from decaying vegetation. That was strange and exciting. So it was to walk on the muddy shore to look at the chunks of fossil corals in the limestone that eroded out along the river.

One of the curious problems about memory is that our continuous memory starts relatively late as children. Even then, "continuous" memory is highly episodic and tied to striking events. I can't tell you what I had for dinner Friday two weeks ago, but I remember clearly the look on the face of the best man on my wedding day in 1965, when he thought he had lost the ring. Still, there are vivid earlier flashes that endure. My earliest datable memory is of being awakened in the middle of the night in an earthquake and rushed outside by my parents. I don't remember the earthquake itself at all, but I do remember being snatched out of bed. I must have been attuned to my parents' emotional state and the haste with which I was bundled out into the dark that night to remember anything about the incident. I recently looked it up, and found that this was the well-known Cornwall-Massena magnitude 5.8 earthquake that took place in September 1944. So I was nearly three. As we stood out in the dark, my father said that our chimney was damaged. I remember looking up, puzzled by what he meant. I could see perfectly well that the chimney was still there. Beth and I stopped in Cornwall in 1996 for a few minutes and drove slowly down 4th Street scouting for the house. My memories of place were sound. It was the same modest brick house, the front porch, the shady setting essentially unchanged. The steel arches of the bridge to Massena, New York, just across the St. Lawrence River were visible from the west end of our old block. A few blocks to the east still stood St. Columban's Catholic Church, which my mother attended, a massive stone building with a startlingly silver steeple – the edges of my former world.

I, of course, had no idea of it when I was a child there, but Cornwall had its own long, quirky history. It was founded in the late eighteenth century, but curiously not by migratory Cornishmen finding roots in the New World. The town fathers were discontented Americans, colonial loyalists who opposed the American Revolution. Our school textbooks skim over the fact that about 20 percent of Americans during the Revolution strongly supported remaining with the British Crown. Several thousand even fought in British military units. A number of loyalist Americans found it to their taste, and certainly helpful in avoiding the hangman, to move north across the Saint Lawrence River to Canada. The king granted them land in Nova Scotia and Ontario. One such group of former soldiers from an outfit called the United Empire Loyalists of the First Battalion of the King's Royal Regiment of New York, accompanied by members of the Eighty-fourth Royal Highland Emigrants, founded Cornwall in 1784. Cornwall later would be on the front lines in the War of 1812 when American forces attempted to conquer the St. Lawrence Valley. The Canadians repelled the invasion at the nearby battle of Crysler's Farm. On the odd Sunday, I would watch groups of men dressed in kilt and sporran march down 4th Street playing bagpipes as they disappeared into the distance – maybe descendants of that founding heritage of the Royal Highland Emigrants, or maybe just a few nostalgic Scotsmen bucking us up with the shrill of the pipes.

Despite the horrors of the concentration camps revealed at the end of World War II, older prejudices lived on in Cornwall as they do everywhere. I had my first puzzling inkling of such things one sunny summer day we went for a drive, with me in my accustomed niche lying on the shelf under the rear window (apparently safe in those pre-seatbelt days). We pulled off the road and drove down a track to a much-anticipated swimming beach on the St. Lawrence River. The beach looked inviting, but my parents refused to stay because there was a sign on a tree at the entrance that read Gentiles Only. I remember the scene because of my parent's reaction, but without comprehending what they said. Children readily notice emotional color. Meaning takes longer.

My recollection of the short northern summer is of exuberance. Summer with my mother mostly centered on the water. She was a graceful swimmer, not a naturalist, and had grown up with lakes. In those summers, she took my sister and me swimming at a local private park on the banks of a clear, slow stream that fed the St. Lawrence. I learned to not fear the water and to swim well enough to stay afloat and make progress from point A to point B. But as I discovered later when I took swimming lessons, I had more splash than style. There was a battered wooden dock from which to leap into the limpid depths. There were water lilies and tadpoles. There was a rowboat too big for me to handle, although I stubbornly tried and ended up stranded in the weeds a few feet away from where I started. My favorite Canadian flower, the bladder campion, with its strange balloon-like blooms, grew among the weeds at the edge of the water. Someone showed me that if you pinched the flowers just right, they would make a satisfactory pop.

What else could anyone want of a seemingly endless summer? Well, of course, in Canada there was no endless summer. Winter came soon enough in all its rigorous glories, and Ontario winter was the real thing. The ground became iron hard. The air was clear, with tiny snowflakes glittering in the sunlight, and then it snowed, and snowed some more, but fortunately not so much that we needed special words for the different kinds of snow. The road crews would create mounds of freshly plowed snow along the street. We children saw them as igloo opportunities and dug cavities into them. Luckily, we didn't manage to entomb ourselves in any snow mausoleums. My mother bought me a pair of kid-sized cross-country skies that I practiced with on in a large empty lot behind our house. When I discovered that I could walk across the snow faster and with less effort than I expended with the skies, I lost any ambition of being a winter athlete. I learned in first grade that ice had its unexpected curiosities. One bitterly cold afternoon, I was walking home along the icy sidewalk with a friend. Up on the hill by the neighborhood corner store we found a dead little dog, a white terrier with all four legs fully extended. He was frozen solid. We pushed him with a stick, and his rock-hard corpse clattered accelerating down the slope of the sidewalk, ice on ice. It was one of those peculiarly disquieting experiences. I had no adult sentiments about the dog, more a lack of feeling because I really could not connect him in any way to my sense of a living dog.

Then my sister Mimi arrived in May 1946. She was named after the wife of my father's friend and mentor Herman Mark. I'm embarrassed to admit that until not long ago I thought she was named for the tragic heroine in La Bohème, a favorite opera of my father's. She had to be baptized Emma Margaret because there was no Saint Mimi in the Catholic Church catalog of names permitted at baptism. The real source of her name is better. I didn't inherit a taste for opera. The plumy portentous voices of the weekend radio opera announcers my father listened to on Saturdays put me off. My mother must have prepared me well for the birth of a new sibling. My reaction to hearing that I had a new sister was eager anticipation of a new ready-to-go playmate. When Mimi arrived home a few days later in a cradle, she seemed awfully small and passive. I put a little toy on her. She didn't do anything. She just lay there. I decided she was useless and for the next few years pretty much ignored her. My brother Bob (Robert Frederick) was born in March 1949. You might ask about sibling rivalry. There wasn't any. I was effectively an only child, and jealousy was not much of an option with that spread of birth dates. It might have been satisfying to exercise some rivalry with sibs closer in age, but that was never realistically on offer. Amazingly, a bit later in life, both my sister and brother ended up being interesting people and wonderful friends. By then it was too late for much sibling rivalry, and my brother had become much taller and more athletic than I was. As a teenager, he gave me his hiking boots as he outgrew them. He grew so fast that the boots didn't even have time to get scuffed before he'd pass them on to me.

He is a Brontosaur:
nine bones and six hundred barrels of plaster of Paris.

Mark Twain

THREE

An Age of Dinosaurs

WE LEFT CANADA FOR PITTSBURGH, A MYSTERIOUS CITY IN Pennsylvania, during the fall of 1949. I know this move was an enormous break in the lives of my parents, hopeful for my father, wrenching for my mother. The trip was just a big adventure into the unknown for me, a train journey to the faraway exotic South. Rail service was efficient and comfortable in those days, with sleepers, dining cars, and authoritative conductors wearing neat blue uniforms. There were lots of windows to gaze out from. The trip was long, and not understanding just how near the equator we were headed, I watched for hours in hopes of seeing exotic creatures by trackside as we crossed into tropical Pennsylvania. Despite my hopes, I was to be disappointed by the scarcity of coiled rattlesnakes and waving palms – but not by Pittsburgh. How could it fail to satisfy? I had never seen a city before. We lived for a couple of months in the Webster Hall Hotel just across the street from Mellon Institute, where my father's research lab was then located. At night the horizons were lit a lurid orange by the blast furnaces. The glow of the furnaces would fade out to extinction in Pittsburgh by the 1980s, and the steel industry would follow. Best of all, our first temporary home was also just two blocks from the Carnegie Museum with its wonderful gallery of dinosaurs. My first visit to that vast, gloomy exhibit hall was unforgettable. I was eight and had never been in such a cavern. The hall contained towering chocolate-colored skeletons, monsters like nothing living today, standing silent, mouths armed with impressive teeth, leg bones the size of trees. Like most children, I was enraptured by dinosaurs. Naturally, I knew none of the scientific drama and wonderful megalomania that lay behind those skeletons. The dinosaurs themselves were enough for my eight-year old sensibilities.

The first dinosaur rush in the American West began in the 1870s and by the 1890s had led to a public explosion of the bitter feud between two quarrelsome paleontologists, Othniel C. Marsh at Yale and Edward Drinker Cope based in Philadelphia. Each passionately strove to be master of all dinosaur bones in the American Old West, a striving equaled only by the passion of their hatred for each other. At the turn of the century, a second round to the glory days of dinosaur discovery would follow as Pittsburgh's Carnegie Museum, New York's American Museum of Natural History, and Chicago's Field Museum vied for spectacular specimens to put on display. The great dinosaur skeletons were the result of extraordinarily arduous years of field work by collectors who worked under a blistering Sun, enduring blowing dust and weeks of monotonous food. They excavated mountains of rock and dirt by hand, then transported tons of fossil bones to the railroad by horse and wagon. The task of cleaning and assembly fell to the museum staff. No single skeleton was completely preserved, so mounted skeletons had to be supplemented for exhibition with parts of other individuals. In addition, the art of mounting an enormous skeleton in standing posture on strong but unobtrusive iron supports was just being invented.

The costs of prospecting, as well as the collection, preparation, and mounting of these skeletons, at Carnegie Museum were funded by the notorious Pittsburgh robber baron and steel magnate Andrew Carnegie, as was the cost of an exhibit hall large enough to hold reconstructed skeletons twice the length of an average forty-foot-long school bus. Carnegie in his later years accepted the responsibilities of his riches and became a philanthropist who generously funded museums, universities, research institutions, and public libraries. There was enormous competition between museums at the time for these new crowd pleasers, and Carnegie wanted his museum to have the biggest of all. As one of the wealthiest Americans ever, he had the money and will to make it happen. The great Jurassic classics, *Apatosaurus* (a massive creature far better known to the world as *Brontosaurus*), *Diplodocus*, and *Camarosaurus* were revealed to all, with their huge bodies, small heads, and long necks and tails. Carnegie was so taken by *Diplodocus carnegii*, named for him in 1901, that he enthusiastically funded yet more dinosaur excavations. The museum scientists who enjoyed his patronage had insightfully named the species in his honor. They repeated the gambit, in 1915 naming *Apatosaurus louisae* after Carnegie's wife. Carnegie ordered casts made of each bone of the *Diplodocus* skeleton. Several exact replica skeletons were used with ego-boosting boasting rights and much hoopla as gifts to museums in other major countries. Only something that large could represent Carnegie's enthusiasm for American paleontology and his self-esteem. I've seen some of these replicas mounted in the dinosaur halls of natural history museums in Berlin, London, and Paris.

There was a spasm of paleontological embarrassment years later when it was discovered that the *Apatosaurus* had been unknowingly mounted with the wrong head at the end of its neck. The bones of an otherwise complete skeleton had been found minus a head and a guess was made so the skeleton could be exhibited. Alas, it would later turn out to be a head from the wrong sauropod, *Camarosaurus*. The chance had to be taken. Who would want to be ridiculed for putting the skeleton of the largest animal on Earth on display without its head? A potentially wrong head was better than no head at all – these are the tough demands of show business. The bad guess about which head to use had been made by O. C. Marsh of Yale, who had discovered *Apatosaurus* in 1879. The actual brontosaur skull was turned up a century later in 1981 by a Carnegie Museum team. While the story sounds ridiculous, the problem arose because although the massive bones of the legs and vertebrae easily survived to become fossilized, the heads of sauropods were lightly built and fragile, and thus rarely preserved.

My favorite dinosaur was the incredibly bizarre *Stegosaurus,* its tiny head mounted on a great bulbous body whose back was ridged by a double row of triangular plates and terminated by a dangerous looking tail surmounted by a double row of sharp spikes. This was an animal too silly to have been made up. There also were some Cretaceous wonders such as the immense three-horned skull of a *Triceratops,* and the chamber was dominated by the skeleton of a *Tyrannosaurus rex,* backed up by a life-sized portrait of a menacing living *T. rex* as imagined in 1949 by Ottmar von Fuehrer, Carnegie Museum's resident artist. The museum's *T. rex* was excavated in Montana in 1905 by the legendary paleontologist

Barnum Brown of the American Museum of Natural History. This astonishing skeleton, as the first of its species ever discovered, became the type specimen to which the name *T. rex* was first applied. In one of those twists of fate, it was sold to the Carnegie Museum at the start of World War II because of fears that New York might be bombed.

In those days when I first met dinosaurs, they were assumed to have been slow reptilian brutes, with their mountains of cold-blooded flesh buoyed up by the water of conveniently available tepid lakes brim full of tender dinosaur-nourishing vegetation. Carnivorous dinosaurs were painted correctly as bipeds, but unwieldy, with their tails firmly planted on the ground. In those older reconstructions these dinosaurs were universally shown poised as passive reptilian stools with three "legs" made up of the two gigantic feet mounting vicious talons plus a massive serpent's tail. In the mural the *T. rex* was shown fiercely resting in this peculiar way. Now we understand from their anatomy and from fossil dinosaur track ways that these animals were active bipeds. Their tails were not slowly dragged behind them but carried horizontally well above the ground. The rigid tails balanced their body weight as a horizontal beam pivoted on their pelvises.

The old exhibit space I loved as a child was closed in 2005, and after a two year period of demolition and reconstruction has been replaced by a dynamic and scientifically updated exhibit of dinosaurs in active poses and accompanied by new murals. The *T. rex* painting of my youth is gone, replaced with a better-informed vision of sprightly hot-blooded dinosaurs walking erect and alert, no more slow reptilian tail draggers. More than that, the entire space devoted to dinosaurs has been greatly expanded by removals of walls and floors to create a vast room proportioned for dinosaurs. Large skylights were built to give the exhibit space natural light and life. The linear arrangement of static dead skeletons along the length of the old exhibit hall was eliminated. Skeletons are now mounted in lively poses in the contexts of their ecosystems. The display exhibits dinosaurs as active animals and shows how their skeletons functioned. Two *T. rex* skeletons square off over the carcass of an unfortunate duckbill. It is a frozen moment before two five-ton beasts collide. The space is divided into the time periods that make up the Age of Dinosaurs – the Triassic, Jurassic, and Cretaceous. All life evolves through time, and dinosaurs evolved with exuberance.

The message of evolution leaps from the exhibits. A series of skeletons and a grotto of skulls show the evolution of the horned dinosaurs, leading from no frills and horns to the final flowering of baroque head frills and massive horns in the later species. Evolutionary biologists recognize that this play on a common structural theme is true today as well in dinosaur evolution. These horns were not necessarily weapons to battle *T. rex* with, as is so often portrayed. Horns, antlers, and many other features have evolved as ways for males to impress each other and attract females. All the diverse shapes of horns of African antelopes, for example, represent sexual selection on horn shape to establish dominance and lure prospective mates and suggest how the hardware of horned dinosaurs may have functioned in display. Dinosaurs may well have been flashy lovers.

The great brontosaurs have been forever lifted out from their fictive warm ponds and tender water plants. They have been liberated to walk on dry land supporting their own great weight and to eat the tough needles of conifers. We even know now from spectacular discoveries of feathered dinosaurs in China that some (likely all) bipedal dinosaurs had feathers like those of birds. Feathered dinosaurs and birds flock together. The evolution of feathers and birds is presented in the new exhibit as a self-contained display of fossils of these feathered dinosaurs and their reconstructed forms. Nothing is stranger to the eye (at least my eye, having grown up with entirely reptilian version of dinosaurs) than a dinosaur prancing around clad in gaudy feathers. Yet they were hot-blooded exhibitionists that new discoveries have shown us were as colored, lively, noisy, aggressive, and violent as the living dinosaurs, the birds. Their skeletons also show that as in living birds, air sacs allowed the single-directional flow of air through their lungs. This arrangement is far more efficient than the in-and-out breathing system mammals possess. The bird system allows for extraordinary high metabolic rates. This setup in dinosaurs indicates their metabolic levels were birdlike, because the air sac system would be needed only by warm-blooded animals and not cold-blooded ones, which have lower oxygen consumption levels.

I can't help but cheer for an exhibit that so imaginatively and with such intelligent use of aesthetics presents the best science, but it is a pity to see all traces of the great original museum exhibits lost as museums modernize their displays to include new discoveries. The discarded old exhibits are part of the history of the evolution of scientific interpretation and public education. Viewing of these old icons and the concepts they illustrated helps make the revolution of scientific thought shown in the new displays more explicable. I'd like to see a display about the evolution of paleontology and evolution exhibits in museums. Our ideas have changed, and so have the images we have used to show off the science along with the bones. Ancient bones can only fully speak when they are seen in a context of ideas.

Carnegie Museum was a magnet to me as long as I lived in Pittsburgh. As I grew up I got to know some of the curators there, notably E. R. (Eugene Rudolph) Eller, an invertebrate paleontologist who helped me identify some of the fossil invertebrates (mostly the shells of clam-like brachiopods and horn corals) that I had found in a hillside. Eller suggested guidebooks to the geology and fossils around Pittsburgh. He and other curators were enormously generous in sharing their time with interested kids. In high school I was allowed to work as a volunteer at the museum. I was assigned to what was then called the Section of Insects and Spiders under its chief curator, George Wallace. I had a rudimentary knowledge of insects and so was turned loose to sit at a long, varnished wood table behind a tray of pinned insects with orders to sort them to their most basic groups and to hand label each specimen in old insect collections. One collection included several drawers from pre–World War II Manchuria. There, among the long dead insects, I first met the enormous fighting crickets that gamblers pitted against each other in combat in little bamboo cages, like miniature, lethal fighting cocks. I also encountered in those collections strange reminders of the tumultuous history of the twentieth century. During the 1930s Manchuria was occupied by troops of Imperial Japan and was named by them Manchukuo. Some of the faded bulk labels I saw carried the original name Manchuria; others carried the Japanese name. It turns out that the old labels reflected an old, much-felt political division among the staff entomologists of the time. The sympathies of these now long-departed curators were battled out in their specimen tags and then forgotten.

There is a distinct museum culture, and I learned a lot of its peculiarities from conversations in the lab and at the long staff table in the cafeteria. There the curators gathered faithfully each day for morning tea, lunch, and afternoon tea. I sat and listened. Among them was another kindly man, O. E. (Otto Emery) Jennings, a paleobotanist, still working in his eighties, who identified many of the fossil plants I had found in the shales from the spoil heap of one of the coal mines and brought in to show him. The 300-million-year-old remains from the coal-age plants of Pittsburgh were spectacular and easily collected – shiny black fern leaves precisely engraved on gray shale. Petrified tree trunks adorned with diamond-shaped leaf scars recorded tree-sized mosses gone mad with ambition. Oval fruit, the fronds of tree ferns, and vine-like forms spoke of the plant diversity of the coal-age jungle and its gloomy wet understory. Occasionally an animal fossil turned up. I once found a dime-sized fossil horseshoe crab, *Euproops,* which seems to have been a semiterrestrial inhabitant of the vast green coal swamps that covered much of the Northern Hemisphere. He shared that world with crocodile-sized salamanders and overblown six-foot-long centipedes and eventually ended up in a drawer in the Carnegie Museum collections.

Each day at the lunch table, Jennings would systematically slit his orange with a small, sharp pocketknife, lovingly removing an intact fine spiral of peel from one pole to the other, a hypnotic operation. I never saw him break a peel. Life at the museum ran on its own rhythm, and it had its share of Monte Python moments. While I was a volunteer, Harry Clench was a curator in entomology and specialist in the taxonomy of butterflies. He spent hours hunched over a microscope, an entomologist monk peering at the genitalia of hairstreak butterflies. These butterflies were so similar in external looks that this is the only way entomologists could identify them and describe new species. Living butterflies appear to have no trouble sorting themselves out according to species and sex – without need of entomologists. Males and females interact first by sight. Then males release a pheromone, which combined with a spiraling dance in flight decides the female's choice. The weirdness of butterfly sex only begins there. We have learned since my time in the museum that some species of butterflies actually have a light/dark sensing genital eye that tells them when the male has properly "docked."

When resting his eyes from the microscope, Harry would happily expound on natural history or gossip. He was sometimes accompanied by Richard Fox, who had become an associate curator after a career as a professional chess player, physician, and explorer. Fox even had the crisp look of the explorer, but it wasn't a pose. In his earlier days he worked in rain forests talking with local healers and collecting medicinal plants for drug discovery. Tropical butterflies were a bonus on which he later published taxonomic treatises. Fox did research on human-biting mosquitoes in Liberia in West Africa, and he, Clench, and others would eventually publish *The Butterflies of Liberia*. This serious monograph, based on the six thousand butterflies Fox collected in Liberia, has photographs of Fox and his wife posed in the bush with butterfly nets. The Section of Insects and Spiders, with its thousands of glass-topped drawers filled with ranks of pinned insects, was located in mothball-scented glory high in the building under the eaves of immense windows. Fox and Clench would occasionally take furtive shots with a blowgun at pigeons resting on neighboring museum lab windowsills.

Although I would only work as a high school summer intern in the museum, my brief experience gave me a perspective on what organismal biologists did. I didn't develop an interest in being a taxonomist, but I learned that important evolutionary questions lurked implicitly in everything that went on in making biological order out of dead insects carefully mounted on pins. My days in the museum provided one of the big intellectual influences of my life. I've never escaped its draw, nor did I ever lose the lifelong love of museums I gained during that entomology-saturated summer.

Education is what remains after one has forgotten
what one has learned in school.

Albert Einstein

FOUR

A School a Minute

MY MOTHER NEVER REALLY ADAPTED FULLY TO LIFE IN PITTS-burgh or felt completely at home with American customs. She always pined for Quebec and for French-speaking friends. Although she spoke English as well as any native speaker, all her life she would emphasize her origins by occasionally using an outrageously fake French accent or interspersing her conversation with "How do you say it in English?" accompanied by a Gallic shrug. She painted avidly and encouraged me to draw and paint. I enjoyed it without being inhibited by the least sense of angst. I knew I didn't have the talent to contemplate becoming a professional artist. My father took enthusiastically to living in Pittsburgh and thoroughly enjoyed being in the United States, which he found amazingly open and free of onerous restrictions. He told me how liberating it felt to live in a country where everything was permitted unless specifically forbidden, as opposed to the authoritarian system he had grown up in which everything was forbidden unless explicitly permitted. Having come to America as an adult, he did find some of the cultural idioms puzzling, however, and never lost his strong Austrian accent.

In the Pittsburgh years it took a long while for the family to settle down in any one neighborhood. I don't know why. Don't you think I might have asked? But children take the vagaries of life pretty much for granted. My father worked for the same chemical and plastics company, Koppers, the whole time my parents were in Pittsburgh, yet they kept packing everything up and moving. Over about a three-year period we moved twice a year, and I each time had to switch schools, usually during the school year. I suspect that the frequent moves were motivated by my mother's discontent with the neighborhoods we lived in. The results of these moves on my schooling can't have been good. They certainly meant that I had to frequently make new friends, and that took time. Then we'd be off again. From my point of view, one of the best itinerant neighborhoods was only a few blocks from an enormous abandoned quarry featuring a dangerous high cliff and the occasional crashing cascade of rock. The place had an easily subverted fence, and I'd sneak off alone or sometimes with friends to adventures among the heaps of truck bed–sized slabs of tan sandstone – an obviously dangerous place to play in and thus ever enticing. This new neighborhood was to lead to my first and only tour of duty as a member of the Cub Scouts. This scruffy but lively troop had no uniforms beyond the little yellow and blue Cub Scout neckerchief, but we didn't mind. Our den leader lived a few houses down the street from us, and our troop of half a dozen met in his house after dinner. He was an astronomy enthusiast who taught us some basic science, such as the speed of light. I was downright amazed to hear that light even had a speed.

We spent most of our scout meetings doing experiments using oranges and golf balls revolving around a light bulb. We all played the roles of planets around that light bulb to understand rotation and revolution, and why we could only ever see one side of the Moon. We stood outside on the lawn looking at the full Moon and discussed how the craters had formed: Were they meteor impacts or giant extinct volcanoes? We Cub Scouts had heated if not particularly well-informed arguments over which kind of lunar violence we liked best. Planetary scientists favored the idea of volcanoes in those days, and the craters of the Moon spoke of volcanoes of unimaginable size. Now, as a result of the revolutionary studies of pioneering planetary scientist Eugene Shoemaker in the 1960s, lunar craters are known to be the result of asteroid impacts. The Moon is now seen to be a palimpsest of overlapping craters marking the great asteroid bombardment of the inner solar system that took place from about 4 to 3.8 billion years ago. The Earth had its share of those ancient impacts, but erosion and the relentless geologic forces of a living planet erased the old scars long ago. Geologically more recent impacts on Earth have left a record of large craters. One massive impact 65 million years ago, at what is called the K-T boundary, significantly influenced the history of life by bumping off all the dinosaurs save those we call birds.

Some of my schools took students on satisfying fieldtrips. Most of the individual outings have melted into a common blur of happy escapes from class, but one moment from an elementary school trip to the Pittsburgh Zoo still stands out with vivid brightness. Defying the organizing efforts of our cat-herding teachers, our noisy class was walking strung out in an unruly gaggle through the large mammals pavilion. The gaggle came up to the rhinoceros enclosure. The rhino was in. Perhaps he was annoyed, maybe mischievous, maybe oblivious, but whatever his state of mind, he slowly swung his massive rear end around and proceeded to pee on the kids. A flood of steaming yellow rhino urine hosed through the bars. Everyone close to the action squealed and scattered. You don't see that every day, and it was certainly good value for a school class event.

I think that the constant school switching had bad results for my education, because I never really connected with any school or with any coherent yearlong class program. My report cards must have been chronically discouraging to my parents. In first grade I had ignored the whole matter of reading because I couldn't stomach even a moment of reciting lines like "See Spot run" from our inane introductory reading book. Things changed when I discovered more elevated literature – that is, comic books. I quickly became a proficient reader. Captain Marvel and Scrooge McDuck were my perennial favorites, but Captain Marvel, to my great regret, was later to die tragically in a copyright battle with Superman. I had to switch loyalties. Although an avid reader, I remained oblivious to arithmetic, spelling, and penmanship. This last is one of the most soul robbing of elementary school experiences. "Make a nice round *O*." Mine looked more like a potato. "Now sit right there and make eleven thousand more." At least that's the way I heard it. Longhand capital *F* was a capital flop for me. Capital *S* looked like a strangled duck in my hands. My favorite school was one that I attended for half a year in third grade. Here was a creative arts– and science-loving place that I thrived in, but it too was left behind by yet another move to another school.

Ironically, worse was to come with stability, when in sixth grade we finally moved into a house in the suburbs of Penn Township, where we lived for several years. Then I was sent to the Catholic parish school. My mother rejoiced because she thought the sisters offered more discipline and more academic rigor than any public school. Perhaps they did, but unfortunately, I was not at all what the good sisters wanted. I was a boy; I was hardly an attentive student; I was not interested in becoming an altar boy. I was, in fact, not even a believer (a failing that I had the sense to keep to myself through all the required stupefying hours of religious instruction in church and classroom). When I look back, it's hard to blame the sisters. They worked hard during their long days in class. One of my teachers, Sister Melfrida, was frowning, fearfully grumpy, and short-tempered, but it's clear on reflection that she suffered from being elderly and from a painfully bad back. She should not have been forced to be on her feet all day coping with unruly hyperactive seventh graders. Still, a certain unreality ruled the roost.

It is truly amazing what we are taught as children, particularly in religious instruction. To scare us about being blasphemers (we were elementary school children, mind you), one of our religion classes featured a hoary old legend about a bad boy who concealed a host at communion and took it home, where he put the wafer into an ashtray and stuck pins into it. The next morning he awoke to find a little body of Jesus lying among the cigarette butts with drops of blood oozing out of those pinholes. As I recall, the bad boy then either went mad and killed himself or confessed his sin, received absolution, and became an altar boy. I'd like to think that the despair and madness option was the heuristically superior outcome.

We were required to go to confession once a week. This – a terrifying event – was my only interaction with a priest. You sat in the nearly empty church until it was your turn to enter a tiny chamber in the wall, close the door, and kneel in the gloom. A wall separated you from another tiny chamber where the invisible priest waited, a portal to God. He slid back a little door and spoke quietly through a black mesh screen. Ten year olds were asked for a recitation of their sins for the week. Think hard. Okay. "I was mean to my brother twice, Father." I'd feel the panic of lapsing into silence. "I forgot to say my rosary all week." And with that, I had survived another week's confession. I can't believe the priests took such picayune sins very seriously. I remember being let off with a piffle: seven Hail Marys and one Lord's Prayer. The little door snapped shut. I'd recite my penance as fast as possible and flee the church. The whole enterprise was a well-practiced form of mind control intended to make us think of ourselves as unworthy flesh and to grow up childishly dependent on the church. Culture sticks, even in an unbeliever. Thus, many years later, I could feel the power and beauty of evensong in the light of soaring stained glass at York Minster. I also enjoy old time gospel songs.

On the science front, Mother Superior, who taught eighth grade, thought that the fresh onion slices she so lovingly placed on the windowsill each morning caught germs and helped keep us flu free. Once after school I questioned this practice. She took me firmly over to the window and showed me an onion slice that had been sitting on the sill under an open window all day. "See the black specks?" "Yes," I admitted. "They are dead germs," she concluded with absolute Mother Superior finality. That was my first lesson in microbiology. The rest of school science didn't lag far behind. I was once marked wrong for answering that aluminum was an element, not an alloy. Producing the periodic table did no good. "It doesn't say it's not an alloy." Maybe these lessons did little harm because we just didn't absorb most of them permanently. The next year I was relieved to enter a public high school, where I no longer had to put the initials *JMJ* (which meant Jesus, Mary, and Joseph) at the top of my schoolwork and exams.

My mother was a practicing Catholic, and we children were brought up as Catholics. My father was a more interesting case. He too had been brought as a Catholic, but a Viennese one, which is to say, not very. He never tried to influence my beliefs, but he inevitably must have by his plainly nonpious example. At various times when I became a bit older, he told me that he was a pantheist, or someone who equates God with nature. He also said he agreed with deism, the idea that God started the universe but has left it to operate under natural laws since. These two are palpably not identical concepts, but they indicate his sense of a god of nature who is not a tinkerer in creation. He did not have any truck for miracles, and in fact was incensed when a friend of my mother's who was from Mexico City lit church candles in fervent thanks to God that none of her family had been harmed in the 1957 Mexico earthquake. My father thought it was barbaric to believe and to be grateful that God had wantonly killed so many other families but generously spared hers. I wonder if she also lit a candle to Voltaire's Dr. Pangloss, the patron saint of such thinking. Not likely.

My early religious disbelief did not come from any study of history, Scripture, or some unlikely precocious childhood reading of David Hume, who was many years off for me. It was just something inborn. The stories we were told in Catholic school about God casting the rebellious bad angels into hell, or rugged Old Testament characters being eaten and regurgitated unharmed from great fishes, or the Creator of the universe speaking to people from inside burning shrubbery just did not seem convincing or inspiring. I was pleased later in life to discover that I was not alone in this kind of innate doubt. There was Edmund Gosse, who was an early-twentieth-century literary critic and the son of Phillip Gosse, the famous British fundamentalist naturalist and contemporary of Darwin. Edmund records in his short autobiography, *Father and Son,* that he grew up in an atmosphere of extreme religiosity. His parents were both members of the insanely strict Plymouth Brethren. He describes not ever believing, despite loving his parents. Once, when his parents had left him alone in the house, he decided to test God's wrath. He put a chair up onto a table and prayed to it as an idol to see if he would be struck down by lightning as promised. Gosse survived, and the chair returned uncharred to its secular household duties. I never went quite so far as to test the matter experimentally.

It has been suggested that religiosity may have a genetic component, that there may be genes that favor a person's degree of religiosity. Any traits that better an individual's smooth psychological fit into a group should be favored by strong natural selection, because the better a chance of a person's successfully integrating socially, the better the chance of that person's mating and leaving offspring. I'd presume that human evolution has involved genes that bind members of tribal groups in a hostile world. In historical time, religion has had a substantial role in maintaining larger cultural cohesion. My lack of enthusiasm for religious tales and belief pressures as a child may thus have had a partially genetic basis. A mental trait that profoundly helped mold my intellectual growth and structure may have no more virtue than my eye color, a genetically linked part of my physical makeup.

It was a relief to finally graduate to the larger and more liberated life of a public high school. Going to the Catholic school had made one contribution, however. Because of all the moving my parents had done, I had made no permanent friends at my earlier schools. This changed. I made two friends during my interment among the nuns, and they remained close throughout high school. One was Jim Weir, with whom I later used to frequent a nearby abandoned strip mine, where we shot at cans with our .22 rifles and launched erratic black powder rockets. In those lax, happy days teenagers could freely buy ammunition and one-pound containers of black powder at sporting goods stores, making such mischief readily available. My other friend was Joe Danforth. His father was an extroverted businessman-engineer who owned a company that specialized in the recovery of industrial diamonds from the sludge produced by grinding metal using copper wheels studded with tiny diamonds. Thousands of dollars' worth of these diamonds looked like a vial full of powdered glass. We would go to Joe's house and listen to his father's off-limits naughty records and drink Coca-Cola. We were pretty innocent. The songs included some hot numbers whose titles I recall as "He Goes to Church on Sunday, So They Think He's an Honest Man," "Cigareets, Whiskey and Wild Wild Women," and the impolitic "Winnipeg Whore." My wife made so much fun of me when I told her the titles that I began to doubt my own memory. I looked them up, and I can say that I had the rare satisfaction of being right. They are all century-old classics of the genre.

Whatever the peculiarities and long, boring times of the school year, summer was a release to life itself. I'm far too slothful to swim much now, but in those far-off days I believed that summer holidays existed for the swimming pool. It was a grievous disappointment whenever there was a polio outbreak, because then my parents would not let me go swimming at a public pool. No one I knew had gotten polio, so polio outbreaks seemed like implausible events of little relevance to me. This is the same separation of the immediate from the remote future that negates any influence of smoking-prevention classes in school. If you don't actually see someone drop dead from blackened lungs in English class, why worry? I had no idea of the terrible fear parents suffered every summer before the

Salk polio vaccine became available. Much later, when I became a parent and I realized what patients confined to an iron lung for life endured, I knew how worried I would have been.

Eventually, the world of water beckoned further. Once I had learned about snorkeling and scuba, I saved money to buy a mask, flippers, and snorkel to see the underwater world of Jacques Cousteau. The possibilities offered by a bland blue swimming pool bottom were sadly not really of Cousteau caliber, but lakes were another matter. Canadian lakes like Lac Souris were best, because the water was green-glass clear and there was much to see below the surface. There were occasional rarities like pearly mussel shells to dive for in the open lake, but marshy areas were best. The waving fronds of water plants offered a strange world full of catfish, perch, insect larvae, snails, and tadpoles. One peculiar sinuously swimming green and yellow creature I caught underwater proved to be a female leech with dozens of young attached. When these little ones started to shift their allegiance in an overly intimate way by attaching to my fingers, I took a rapid boat break.

Beth tells me that at the same age she lived perched on back of a horse whenever she could during the summers she spent at her grandparents' farms in Indiana. She rode bareback and even read novels sitting astride her sturdy farm steed while it grazed placidly under the shade of a box elder tree. I was a suburban child, and so I rode a bicycle when I wasn't swimming. I conceived of riding horses as a remote, difficult, and rare skill that the Lone Ranger used when trouble was brewing, or at sunset when he was departing. Yes, the Lone Ranger, his sidekick Tonto, his trademark black eye mask, the silver bullets, his unerring accuracy in shooting the six gun right out of the hands of evil malefactors, the hokey stilted dialogue. I was a faithful fan, glued to a television as big as refrigerator, my face lit by the glow of a screen as big as a postcard. In the one episode that left my father the chemist hooting, the Lone Ranger kneels by a water hole surrounded by dead cattle, sticks his finger into the murky pond, tastes it and says "Tonto, this water is poisoned."

In 1963, after I had left home and was about to graduate from college, my parents made the last big family move. This time their destination was Pullman, Washington. I think moving was driven by my mother's desire to leave Pittsburgh, and by my father's discouragement at seeing his company reduce funding for its research efforts. Fortunately, the romantic West proved to be successful for both of them. They immensely enjoyed living in the Snake River country, and my mother was able to teach French part time at Washington State University. She also finally had a nearby social group of neighbors, mostly other wives of college faculty, whom she felt comfortable with. My father, for his part, was to achieve a striking (perhaps unique in the history of the far West) sartorial effect on weekends. There he faced the western sunset dressed in Austrian lederhosen (leather shorts) combined with sandals and socks, a cowboy hat, and silver bolo tie.

My father suffered a stroke in the mid-1970s. Although he recovered from most of its physical effects, he lost the one activity he cherished most of all. He could no longer read effectively, cutting him off painfully from his greatest pleasure and much of his intellectual life. He felt hugely diminished and became a passive observer where he had once been an active presence, always ready to discuss events and ideas with us. My father died in 1981, just before we were to visit for Christmas. My mother was a different character. She never gave up hope and remained vigorous, gregarious, and lively until shortly before she died in 1997. A few months after my father died, she sold her house, left the Olympic coast she loved, and went to live with my sister's family in Virginia. She existed for her new grandson; she continued to paint and sculpt. Astonishingly to me, she even took up golf. My parents are buried in the town where they met, Shawinigan. They rest next to my mother's parents.

When you have seen one ant, one bird, one tree,
you have not seen them all.

Edward O. Wilson

FIVE

In the Natural World

CREATURES

I was an inveterate naturalist. Each year I anxiously awaited the return of spring (and, truthfully, the end of the time-crawling endless school year). I felt a strong curiosity and an intense attraction for the look and feel of natural forms and creatures, the stranger the better. At various times my interests settled on hunting salamanders, insects, turtles, snakes, and fossils in the forested hills near our house. I had read that snakes had no eyelids, so I had to look a snake in the eye. Sure enough, their eyes are covered by the clear window of a single modified scale and can't be closed even in sleep. All snakes are carnivores. I kept snakes and watched them feed using independently attached lower jaws armed with sharp, curved teeth. A snake engulfs its prey by walking each jaw alternately down its victim's body, and there is no escape once a snake begins to swallow. It happened to me. I was handling a middling sized garter snake, about eighteen inches long and about as thick as my index finger. It bit the tip of that finger and held on. This posed a quandary to both of us. The snake couldn't let go because of its recurved teeth, so it began to work its jaws up my finger, committing itself to swallowing a nearly full-sized human – a new frontier for a garter snake. I carefully disengaged its independently movable lower jaws, and slid my finger free without hurting the snake. Fortunately, its upper jaws hadn't secured much of a hold because my fingernail was in the way. I got to keep a few tiny punctures as souvenirs.

This story suggests something of the vagaries of natural selection. The feeding mechanism of snakes is highly evolved and has served them well, yet evolution can't look ahead to situations in which holding on and swallowing might be fatal to an individual snake. I once stumbled upon a strange-looking dead snake. The back half of another snake protruded from its mouth. It had tried to swallow a snake of almost equal size. I pulled the swallowee out. Its head had been digested, but not fast enough to save the consumer from suffocating. There are also documented cases (at least in captivity) of hungry snakes that mistakenly attack their own tails. Once their teeth have fastened on they can only move forward, which means eating their own bodies to the point where the body can't bend any tighter – or having the good fortune of being rescued by a veterinarian.

Liking drama, I watched orb weaver spiders capture insects in their webs, wind them in silk, and then inject them with venom. They later inject a cocktail of digestive enzymes and feed leisurely on the liquefied insides of the insect. More primitive spiders bite first and then wrap dinner up to marinate. On slow days I'd help out by providing the prey to see how different insects fared. Large beetles and grasshoppers simply muscled their way out by breaking the web. The biggest question we kids puzzled over was how the spiders themselves could walk around on their webs without becoming stuck. This conundrum was well solved by studies showing that spiders make different kinds of silk. Radial strands, made first in building an orb web, are not sticky. The spiral silk put down afterward is the sticky trap for insects. The spiders simply keep their feet off the sticky strands. It's the females that make the spectacular silken death traps. By end of summer they have become fat and ready to mate. That the words "spider" and "sex" could be associated in one sentence had not occurred to me. It was only years later that I first saw a little eight-legged romantic male spider woo a big greenish brown female *Neoscona* orb weaver who built her web each day from the eaves of our garage. The scrawny males timidly hang out at the edge of a web and pluck the radial strands hoping to convince the female to accept them as sex partners rather than dinner. The job, once an ardent male is accepted to her lair, is to pass on his genes by successfully mating with his ladylove before she changes her mind. It's often not really a choice. In some species, successful male spiders go to their reward as postcoital snacks for their unsentimental mates.

One year's enticement was salamanders. These small creatures were common in Pennsylvania woodlands if one was willing to spend time turning over rocks and rotting logs and poking under the stones in rivulets. The diversity of lungless salamanders was high, as the eastern United States is a hot spot for the evolution of this odd family of salamanders, the group to which most of the North American species belong. The biomass of salamanders in eastern U.S. deciduous forests might be higher than all other animals, but they are secretive and rarely seen. Some are startlingly beautiful and strange, with species' colors ranging from yellow and green to red, and muted purple. The slimy salamander is a gorgeous jet black with white spots, and if you mess with it, the penalty is to have your fingers slimed by an intense outpouring of defensive mucus that turns black and is only removable by time. The formal name *Plethodon glutinosus* tells all. I didn't lick one to see if it was toxic or bad tasting. I wasn't that much of a naturalist. The most gaudy of all the salamanders in my patch of forest was the damp-loving longtailed salamander, which is golden colored with rows of black spots along its body.

Lungless salamanders are small and live in moist environments, and so they have dispensed with any need for the luxury of lungs. Wet skin does the job just fine by way of gas exchange. They feed on small insects by lightning fast projection of a sticky tongue. Given their numbers, this tongue-in-cheek method makes them some of the most successful predators we have around us. These salamanders are no more than a few inches long, but they are distantly related to much larger species of salamanders. The largest salamander in the world is the five-foot-long Japanese giant salamander. The largest in my world was the eighteen-inch hellbender, which is caught in eastern rivers by less-than-pleased fishermen (it can bite careless fingers quite competently).

Sadly, the wonderful amphibians of the world are rapidly declining. Over a third of the nearly sixty-seven hundred known amphibian species are threatened or going extinct as I write. Declines can be almost instantaneous. For example, David Wake, who studies the evolution and conservation of salamanders, has documented their crash – and it can only be called a crash – at sites in Central America from the 1970s to

2005. The loss of these beautiful creatures also includes the salamanders of the eastern U.S. woodlands. The worldwide causes of the declines vary. A rapidly spreading fungal disease and the local effects of climate change are implicated in some cases. Frustratingly, the causes of other declines are not yet known. In highly developed areas such as the eastern United States, habitat loss is one of the possible causes. Creatures that lived through the extinction of the dinosaurs and have evolved into some of the loveliest of animals are disappearing before our eyes. The price of memory is the knowledge of loss. It is even more poignant when the natural world itself melts around us.

I loved exploring, but it turned out that I also was an incurable collector, so when I volunteered at the Carnegie Museum I started an insect collection. I soon learned there were some logistic problems attached to the greedy accumulation of dead bugs at home. The books showed me how kill, pin, and label them. I had a small bottle of cyanide that I used to make killing jars – I can't believe now that cyanide would be so easily available to a teenager. I soon changed to using the much safer nail polish remover (ethyl acetate). But there was less in the way of practical instruction as to how to keep large specimens from decaying while waiting for them to dry on those pins. There was no way to conceal from my mother what happens in the dreamy heat of summer to those big dead cicadas. "Something smells rotten in here. It's those dead bugs on your dresser." There are ways of removing guts from big insects such as caterpillars and moths, and there are fast ways to dry specimens. The second problem was harder for me to solve. Where do you put them all? A few insects are big, with large wingspans – such as butterflies – but most are small. However, there are ever so many species of them. Just how many is still not known, let alone for the rest of the world's phyla of small creatures.

A number of years ago, an ecologist named Terry Erwin made a startling estimate based on a seemingly straightforward experiment. Erwin wanted to know how many species of beetles might be living in a single large tropical tree. He put out large sheets under a tree in Panama and fogged the tree with a cloud of commercial insecticide. At the end of the day, he folded up the sheets and took his arthropod booty back to the lab. Then the next day he fogged another tree, then another, and another. He spent the next few years in what must have been a profoundly tedious sorting out of his catch. Erwin then made an interesting rough calculation. He noted that each species of tree had its own specialized species of beetles that could only live on that single species of tree. He estimated these at about 163 beetle species. There are 50,000 tropical tree species in the world. Erwin made a rough estimate as to what fraction of arthropods these beetles represented and suggested that there may be as many as 30 million tropical forest arthropod species on the planet. These would include all the other insects, such as ants, as well as spiders, mites, silverfish, millipedes, and pseudoscorpions – most of these species not yet discovered. Seems outrageous doesn't it? A more sophisticated estimate, published in 2010, suggested that Erwin overestimated the number of species, with three million being more likely. Another study in 2011 settled on nearly nine million eukaryotic species (animals, plants, fungi, and protists). These are big numbers, and once we think about the rest of life, we have to start in on the vast unwashed hoard, the horrendous number of species of bacteria being found by gene sequencing in every sample of an environment. If we add in the suspected numbers of kinds of viruses, the lid is likely to blow right off.

Erwin had yet to make his famous estimate when I started my insect collection. Nonetheless, it soon became clear to me that even the miniscule fraction of the insect multitude that hung out in my neighborhood threatened to overwhelm any possibility of identification and make a collection impossible to store without asking my family to move out. I decided to simplify life by collecting dragonflies (and their elegant cousins, the damselflies). These are among the largest and most dramatic of insects and have a history dating back to hawk-sized giants that cruised the swamps of the coal age. There are also only a few hundred species of dragonflies in the entire United States, which seemed like a manageable number, and their biology turned out to be satisfyingly fierce. Adult dragonflies are predators, snatching other insects in flight. The males are territorial. They have brightly colored wings or bodies and have elaborate ritual fights with other males for control of a bit of the shoreline of a pond or stream where they hope to meet similarly inclined females.

In the just course of things, eggs are laid and babies hatch. Baby dragonflies are otherworldly, weird aquatic forms. These larvae are called nymphs, but they are less like coy nymphs than killer aliens on a small underwater scale. The nymphs of the largest species are powerful enough to catch and eat minnows. The nymphs take water into their abdomen from the rear and breath internally that way. They also can expel the water with great force and jet forward like torpedoes. They have keen vision, and, oh yes, their best feature, horror movie jaws. Their jaws are made of what corresponds to the "lower lip" in the adults, the labium. The adults have real jaws, mandibles, but in the nymphs the labium is expanded into a folded structure that's kept tucked up under the body and is about as long when extended. The extensile labium includes at its business end a pair of impressive toothed "jaws." Nymphs hunt from ambush, slowly stalking anything up to their own size. Their keen vision lets them target their prey and shoot out their extendable long lower lip with its immense jaws to grasp their victim. Few prey items get away from that speedy strike.

I borrowed a pond dredge – a sort of sturdy metal basket on a wooden pole – and started catching nymphs from the debris that lies on the bottoms of ponds. I kept the ones I caught in an aquarium and had the pleasure of watching them deliberately stalk and ambush worms and aquatic insects. The nymphs thrived. Some made it to adulthood. Others were ambushed and cannibalized by their brother and sister nymphs. The emergence of the adult is glorious. The full-grown nymph climbs out of the water on a twig and its "skin" begins to split along its top. A dragonfly slowly emerges from the husk of the nymph and, hanging on to the empty skin, extends its body and shimmering wings. Its body hardens, and in a few hours the newly minted dragonfly lifts off for a jolly life of piracy and debauchery.

My collecting led to my first scientific discovery, minor but my own. I found a little blue damselfly of a kind that I hadn't seen before flying at my favorite dragonfly pond. I identified it from a book on dragonflies as being *Enallagma basidens*. It looked like a solid match, but the books on dragonflies said this species was only known in southwestern states. I had my identification checked by an expert. It was an interesting lesson that sometimes the book is wrong. I wrote my first scientific publication (a single paragraph long) on how the range extension of this species had gone as far as Pennsylvania for the journal *Entomological News* in 1960. My discovery was a small fact, but science is not just an accumulation of facts. To become science, observations need to be woven into the testing of hypotheses and integrated into the larger body of data that make up a scientific discipline. I hadn't reached that point.

In my last year in high school, I learned that moth collectors had devised a rich, smelly bait made of beer and molasses that could be put out as a lure. All that remained to enjoy night entomology when I should have been doing homework was to go out into the woods on a "bait trail" I painted on trees by day and revisited by night with a carbide lantern and cyanide jar. There I'd collect the moths that came by for a free drink at the tavern. The most striking were the large *Catocala* moths, with their red or yellow colored underwings. The bright underwings are normally concealed as the moth hangs out on tree bark for the day. But the moth can suddenly flash these "warning colors" if a bird should want to snack on the otherwise dull moth. If the bird is startled even for a moment, the moth has a chance to drop off its perch and disappear. These are favorites for moth collectors, some of whom make crosses between species to get hybrid wing patterns. I became fascinated by the idea that there was active unseen insect life in the midst of winter when the woods looked pretty dead as far as insects were concerned. I tried winter baiting and found that a few modest species of moths came out on winter nights. Among them were uncanny wingless insects. These were the flightless females of the cankerworm moth. They climb trees and attract their flying mates with a pheromone broadcast into the air. Horticulturists wish they wouldn't do that.

I was starting to take part in one of the most unlikely of human activities, science, and the enterprise that arose so rarely in thousands of years of human history. Science began in ancient Greece, but except for a few notable exceptions it never became in ancient times an experimental and technological enterprise. The rise of modern science followed a torturous path after the fall of classical civilization, being preserved by a series of brilliant scholars in the early Islamic world while Europe was still grinding painfully through the Dark Ages. It was further enriched by an opulent tradition of mathematics, science, and technology developed by Arab, Indian, and Chinese scholars. Arab scholars translated the surviving Greek manuscripts and themselves made important new discoveries in astronomy, chemistry, and medicine by use of observation and experimentation to test their hypotheses. The first European rediscovery of ancient philosophy and science via Arab translations was made in the twelfth century of the works of Aristotle. The flood of translation became the driver of the Renaissance.

The scientific revolution that began to accelerate in Europe in the sixteenth century had gotten its boost from the rediscovery of ancient learning but required a new start in which the limitations of the older Greek and Arab scientific ideas were largely replaced by entirely new concepts and experimental approaches. Science also feasted on the discoveries of the Age of Exploration with its disturbing revelations. The discovery of unexplainable peoples and unclassifiable creatures brought home by the explorers of the sixteenth and seventeenth centuries burst the limited boundaries set by the known animals and plants of Europe. Discoveries that would go to the heart of nature's workings would depend on an experimental tradition fostered by the birth of instrument making. Because of the revolutionary invention of the printing press, new discoveries could reach everyone quickly. Books were no longer priceless hand-copied and inaccessible rarities.

The downfall of older ideas began with the heliocentric model of the solar system devised by Copernicus and visualized by Galileo's use of the telescope, which displaced the Earth from the center of the universe. An analogous transformation in medicine began with the publication in 1543 by Vesalius of his revolutionary human anatomy, which overtopped the Galenic tradition in medicine. Roman law had prevented Galen from doing human dissections. He thus substituted monkey cadavers for humans – an amazingly perceptive choice for the second century AD. Galen also was an unusually early experimenter who showed that the brain controls muscle action. Vesalius's revolutionary advances were based on a new source of data, the freedom to perform dissections of human cadavers. Somehow the flood of discoveries inspired the rise of early modern science in Europe, an event unique in human history. The conditions for the ultimate blooming of science in its modern form required a loosening of the power of religious dogma. When religious groups are in absolute control, dissenting from their views generally translates into heresy trials that promise painful outcomes for unbelievers and for the discoverers of inconvenient facts of nature.

The idea that events such as earthquakes and plagues were natural and not supernatural events sent by God or caused by the devil only came in the eighteenth century. Benjamin Franklin showed that lightening was merely on a larger scale the same electricity that could be generated in experiments. This was a major step in understanding that great phenomena could have comprehensible natural rational explanations, that, in fact, they demanded natural explanations. In Franklin's case, lightening rods were the practical outcome of this knowledge. They would displace the common practice of ringing church bells to frighten off spirits at the approach of storms. Reactionaries would claim that the deployment of these simple devices was an impious attempt to deflect divine retribution. Lightning rods were also claimed to cause earthquakes, but even earthquakes, these handy tools of divine displeasure, were to receive scientific attention in the Enlightenment. The modern culminations of the displacement of the young Earth and human-centric view of the universe came from three discoveries made in the nineteenth and twentieth centuries, the immense age of the Earth and life, evolution, and the universality of DNA and the genetic code among all forms of life on Earth. These concepts came to circumscribe most of my scientific universe and research.

Most children are curious about natural objects and living creatures, starting with the desirable tidbits they put in their mouths as infants to taste (and so horrify their mothers as the dead bug moves precisely from fingers to lips). Their curiosity progresses to about everything. But this intense and apparently innate early interest in how the world works tends to fade for most children as they grow up. Inquisitiveness is often actively stamped out by the teen years by the rise of more compelling hormones and the distaste for the compulsory sitting through of those less-than-stirring high school science classes. Yet some remain curious.

I think that there are two chief ways children mature to become scientists. These two paths seem to produce two significantly different flavors of scientists with distinct interests and intellectual styles. What I call Type I scientists are evolved from the kid naturalists who were stimulated early by a love of nature and never quite got over it while they should have been growing up. They like roaming the outdoors, don't find bugs and worms yucky, and can bemuse or wear out their parents by their propensities to bring home bird's nests, butterflies, snakes, rocks, and fossils. Most of these kids don't become professional scientists, but many remain naturalists all their lives and fill the ranks of dedicated birders, amateur paleontologists, and nature photographers. If they eventually do develop into professional scientists, they are likely to end up in organismal biology, marine biology, paleontology, ecology, or evolution. I was one of those. Type II scientists arise in quite another way. They come from the ranks of the bright and curious kids who didn't mess with bugs but along the way are entranced by the intellectual elegance of science by fortunately being exposed to an inspiring high school or college teacher. They are likely to enter what are sometimes thought of as the more respectable branches of science and become physicists, chemists, geneticists, or molecular biologists. Not surprisingly, there is some overlap – even in me. I am a synthesizer and in my career have liked best making the link between previously unconnected disciplines, such as blending evolutionary biology and phylogeny with developmental biology and, most recently, tying microbiology into the study of how fossils form.

EVOLUTION SECRETED

I somehow learned that life had evolved, but not from any biology course in high school. As was typically done, my biology course was taught in ninth grade. The teacher was a balding chunky football and wrestling coach who also had been recruited to teach biology. I was a little nervous at first, but he turned out to be a good teacher and a tolerant and pleasant man. The E-word was never mentioned. We were not taught creationism, just basic biology minus evolution, minus Darwin. Our teacher had us each write a term paper and give a brief presentation. I did mine on the evolution of snakes and lizards. No one in class swooned at the mention of evolution, because it wasn't an issue. Evolution was nearly invisible in schools in the 1950s. After the Scopes Trial of 1925, textbook publishers had simply downplayed the troublesome concept to avoid further challenges and loss of profits. We had good chemistry and physics classes, but biology was weaker, because as a "soft" science it was taught in ninth grade, before students had absorbed much if any science. In my memory, there was little mention that I can remember of basic evolutionary topics: the meaning of fossils, the progress of life through time, how we know about the age of the Earth – they were illegitimate children.

As I wrote this account, I became curious about what our textbook actually contained in the way of evolution in 1956, when I took high school biology. I remember the book we used, by Moon, Mann, and Otto, because at the time I was amused by the thought of it being authored by one "Moon Man Otto." Remember, I was fourteen when the book was issued to me. I recently ordered a copy of the 1956 edition of Moon, Mann, and Otto's *Modern Biology*, probably the most widely used high school biology textbook at the time. Half the schools of America used this text. My five dollars was well spent. The 1921 edition of its predecessor, Moon's *Biology for Beginners*, had a portrait of Darwin as its frontispiece, but that gaffe was soon cured. The portrait had shuffled off by the 1924 edition. The word "evolution" doesn't appear in the 1956 Moon, Mann, and Otto but was replaced with a truly grotesque neologism, "racial development." A chapter titled "The Changing World of Life" is devoted to the topic that we would call evolution, but it doesn't appear until page 657, as chapter 51. Students are unlikely to have ever reached a chapter this far back in the book. It hardly matters, as it was a paltry and muddled sort of evolution in Moon, Mann, and Otto's timid hands. Ironically, biology textbooks authored for Catholic schools in 1956 contained the word "evolution" and covered the topic far better than our craven text.

Moon et al. offer up a thoroughly schizophrenic view of Darwin's science. First came a reasonable if way too short summary of natural selection, but that was followed by the disclaimer that selection leaves many questions unanswered; "one hand giveth, the other taketh away" was the safe way to go. This is one of several surrenders to creationists. The first surrender in fact comes in the first line of the chapter, which reasonably asks, "How old is the Earth?" Immediately, though, the authors then shrug the whole issue off. "It probably doesn't matter too much, and certainly will never be settled." To their credit Moon et al. do go on to present an illustrated geological time scale and acknowledged a billion years plus for the age of the Earth. The best of the weaseling is saved for last, at the very end of "The Changing World of Life" chapter. There is a final little section called "Science and Religion." It might have well have been called "Science Equals Religion."

In that last section, Moon, Mann, and Otto used two peculiar ad hoc quotations to reconcile laboratory and church. One is by W. W. Keen, the pioneer American brain surgeon, who saw God's hand in the mechanisms of nature. He was an outstanding surgeon, but his opinions on evolution are as extraneous as mine on neurosurgery. The other curious quotation is by Darwin's close friend and vocal evolutionist Thomas Henry Huxley, the coiner of the term "agnostic." The quotation is used egregiously out of context. The cited line comes from a long rambling letter Huxley wrote in 1860 to his friend, clergyman and Darwinist Charles Kingsley: "Science seems to me to teach in the highest and strongest manner the great truth which is embodied in Christian conception of entire surrender to the will of God." But that was a gross and misleading use of Huxley's words. Huxley's four-year-old son had died of scarlet fever only a few days before, and in despair he was responding to Kingsley's letter of condolence. Moon et al. made it into a nice bow to creationists by inserting it as an out-of-context bit of Christian theology into a public school textbook. Huxley, in fact, struggled most of his long career to replace the prevailing religious concept of nature by a scientific one. A more characteristic quotation might be Huxley in the *Westminster Review* in 1860: "Extinguished theologians lie about the cradle of every science." Yes, but it wouldn't have been the smarmy and safe accommodation to religious sentiments that Moon et al. wanted to make.

The lack of the most basic information offered about evolution by the public high schools of the period would make students leaving high school easy prey to the claims of creationists about missing intermediate forms in the fossil record or their asserted young age of the Earth. It is little wonder, with the frenzy of self-censorship by authors and publishers, that the creationist circus largely had fallen silent. Their work was done for them. Only one author of the post Scopes Trial period through the 1950s, Ella Thea Smith, seems to have had the integrity to honestly discuss evolution. Her text *Exploring Biology* included substantial treatment of evolutionary biology, including human evolution. Astonishingly, the book had sales reaching to as high as 25 percent of the market. Biology textbook authors rediscovered their enthusiasm for evolution during the science boom of the 1960s, with the publication of the National Science Foundation sponsored BSCS (Biological Sciences Curriculum Study) textbooks.

After that, the evolution content of texts returned to bobbing and dancing in tune with the political winds. Publishers and authors again shied away from evolution under creationist pressure during the 1970s and 1980s, the era of equal time for "creation science." When the courts drove a stake into that oxymoronic concept, publishers in the early 1990s not surprisingly found that courage and profits could align. Now the best current textbooks treat evolution thoroughly and well, but the teaching of evolution in modern American public high schools is spotty because of local opposition in some communities, and even the illegal actions of creationist public school teachers and school boards. Creationists also publish textbooks intended for the rapidly growing alternate reality of fundamentalist schools and home schoolers.

I was an addicted reader from grade school on – no surprise – at first mostly boy's adventure stuff, like the life of Wild Bill Hickock (which disappointingly ended in the bleak Dakota Territory by his being shot in the back of the head by a fellow saloon gambler). Once I found the books hiding in our school library, I also enjoyed the "orange biographies" published by Bobbs-Merrill that were popular at the time. Some were about the young lives of famous scientists, with inspiring standards like Thomas Edison, Louis Pasteur, and Ben Franklin being prominent fare. More sophisticated biographies and autobiographies followed, and so I learned about Cope and Marsh and their great dinosaur bone feud of the 1870s and 1880s, absorbed the tales of great naturalists, and finally got into the real history of science. Reading about what scientists did was just as important to my thinking of myself as a scientist as were my own rudimentary experiences as a naturalist. The scientists I read about supplied important role models, although the worlds they inhabited and the questions they asked had largely vanished.

Eventually I heard of Darwin's *Origin of Species*. I first tried to read *Origin* when I was a senior in high school. I thought naïvely that it would contain lots of stirring tales of the fossils that I imagined served as the main fonts of information about evolution and thus were surely the noble keystone of Darwin's thought. Nary an *Iguanodon* was to be found. For some inexplicable reason, *Origin* devoted what seemed like an inordinate amount of print and paper to discussions of the breeding of pigeons. I was deflated. It wasn't until many years later when I actually had learned some evolutionary biology that I understood that what Darwin was getting at was the development of a mechanistic theory of evolutionary change. All those bizarre breeds of pigeons that he talked about illustrated the workings of artificial selection, the human analog of the operation of natural selection in nature. The pigeons thus were a crucial part of the long argument that Darwin laid out in *Origin of Species.* Their role was to show how the natural selection he was proposing worked by analogy with the selection practiced by animal breeders on natural variants. By selective breeding it is possible to get certain traits to predominate in domestic stocks. Similarly, a mechanism of natural selection that favored the reproduction of some heritable variants over others could work in nature over long time periods. The best place to start reading Darwin is not with *The Origin of Species,* it's with *The Voyage of the Beagle,* which is arguably the best travel book ever written and opens to us the thinking of young Darwin during the adventures that ultimately led to his life's work.

There were other books that influenced my thinking about science. Two books I particularly remember were ones I read on vacation in Quebec the summer after my freshman year at Penn State. These were *Genesis and Geology* by Charles Gillispie and *The Origin of Life* by the Russian visionary Alexander Oparin. The first book, on the impact of evolution in Victorian Britain, was a revelation on how science had a social context: Not only did science affect culture, but the intellectual ideas current in a culture played a large part in molding scientific thought. My grandmother, who always seemed to be a determined francophone, proved to have plenty of English comprehension when it counted. But then, our grandmothers generally do have more savvy than we are likely to give them credit for. She spotted the book and told my parents that she didn't think it was the sort of thing I should be reading. But she was wrong. *Genesis and Geology* showed me that there were lots more glorious books about the history of science to be read. I was hooked.

Oparin's contribution was to address the mystery of mysteries in a chemical format. His book reinforced the decision I had made to change my major from chemistry to biochemistry in the coming fall. I first learned about the specificities of biochemical components of life like the amino acids and that these could perhaps be formed naturally without life. This idea recently has been borne out by observations that amino acids are present in meteors and comets and have survived in meteorites found on Earth. The arrival of substantial amounts of organic molecules to Earth by collisions with objects from deep space is more than an exotic idea. But Oparin had a more mind-expanding hypothesis. He suggested that the Earth's atmosphere in the distant past was not the oxygen-rich mix we have now but something more like the atmosphere of Jupiter, with methane, water vapor, and ammonia all available to react with the right energy input and produce the chemicals of life. His realization that such a variety of chemical events could take place in the evolution of a planet was a breakthrough in how life and its origins thus could be explained naturally – if we are smart enough. We haven't been that clever yet, but the idea remains powerful. New suggestions about the possible interactions of early life chemistry with geochemical reactions at mid-ocean rifts that produce chemical gradients may help make the questions answerable.

Supposing is good, but finding out is better.

Mark Twain

SIX

Transformations

AFTER ONE MORE FAMILY MOVE, I DID MY LAST YEAR BEFORE college in another new school, Gateway Senior High in the Pittsburgh suburb of Monroeville. As many of my classmates were also new students who had just transferred there with the metastasis of suburban sprawl, no cliques of cool kids had built up and the teachers were good. Perhaps the best was our exceedingly sarcastic English teacher, who supplied just the right attitude for reading about the oddities of the Macbeth family and Julius Caesar's unfortunate misjudgment of his friends. Two of my Gateway friends eventually took doctoral degrees in science. Tom Taylor became a mechanical engineer, and Mary Boesman became an immunologist. She died in 2007. Mary and I both read history and lent each other books. This was the best of my twelve-year run of schools, and I remember it fondly.

But that pleasant year also included the looming matter of where I should go to college. Given what my family could afford, my choices were limited. I could attend a university in Pittsburgh and live at home or attend the state university and live away. Not much of a contest. Despite the good universities around Pittsburgh, living at home would suffocate my becoming independent. After all the adventures of filling in applications, taking College Board exams, being interviewed, and sitting in on a sprinkle of college class lectures, I left for Pennsylvania State University in the fall of 1959. My parents took me to State College and helped me move into my first dorm, but I was anxious for them to leave. I now had to make my own triumphs and my own missteps (ludicrous missteps outnumbered triumphs that first year). Our first assembly as freshman students included an address by the president of Penn State. He told us to look at the person on our left and the person on the right. One of them would be gone before the year's end. Perhaps that encouraging invitation to embarrassed sidelong glances constituted what might be described as a subtle preparation for life.

My mother, who had tolerated years of pet salamanders, turtles, snakes, and lizards, kindly agreed to care for the last of the reptile host after I left. That was Ferdinand, my South American green iguana, who couldn't be just released back into nature in Pennsylvania like the local reptiles I kept. Unfortunately, she didn't know that reptiles are highly sensitive to insecticide, and so she sprayed his cage to rid it of fruit flies. It was a lethal neurotoxin for Ferdinand the iguana too. College was my first experience living away from home, and my first year was tough, because I was homesick and because I was not really enthralled by most of my classes. I was enrolled as a major in chemistry. This choice may seem surprising considering my natural history interests and my formative experiences at Carnegie Museum. However, my father was a chemist and persuaded me that doing biology required learning hard science as well. I had a strong liking for chemistry itself, but not for much of the rest of the requirements of the program, especially the courses I had no aptitude for: math and German. I enjoyed the Navy Reserve Officer Training Corps (NROTC) as something completely different – even the weekly drilling in brilliantly shined shoes and pressed uniforms. We marched with shouldered rifles across a well-mowed campus lawn accompanied by a brass band. The U.S. Navy is traditional in its ways. Our drill rifle was the bolt-action model 1903 Springfield, designed at the end of the nineteenth century.

On the suggestion of my Carnegie Museum friends, I got in touch with George and Alice Beatty, two entomologists who lived a short distance away in a tiny village, Lemont, at the foot of Nittany Mountain, a long ridge that looms over State College. George was an autodidact with a photographic memory and a spectacular library of books on natural history. He was simultaneously charming, expansive, and difficult. Alice had a lilting Texas accent. She was generous and ineffectual. One was decisive, one was indecisive, and so they made good partners. George came from the Philadelphia suburbs and lived on an apparently quite satisfactory family inheritance. Almost unheard of in the twentieth century, he worked in the manner of a nineteenth-century gentleman scientist in his house on his own funds, generating a vast private dragonfly collection. Despite his enormous talent, George completed little research, but he taught me a treasure of natural history lore and the basic skills of wildlife photography. He and Alice provided a sort of home away from home and took me out on frequent weekend dragonfly collecting trips in the little known wild places in the mountains – a wonderful break from campus. The mountaintop peat bogs and the clear streams of the Appalachians were a world of unique environments with unusual plants and insects. That's where I saw my first insect-eating pitcher plants.

George and Alice introduced me to another lifelong friend, Hal White, then a lanky high school junior with an amazing pitching arm and a craze for collecting insects. We waded a lot of streams together looking for dragonflies. Once we took a long drive to Tioga County with George to see a quaking peat bog in remote northern Pennsylvania. It was a hot day, and we stopped at a general store along the road near the bog to buy soft drinks. While Hal and I were picking out Cokes and snacks, we noticed George leaning over the battered wood counter in earnest conversation with the clerk. As we walked back to the car, we asked George what that was all about. George was wearing an old tan army shirt, and the shopkeeper thought he was a warden from the juvenile corrections facility just up the road. He also thought that we were George's charges. In a strange rural town, this was not at all a happy turn for a conversation about us to take. At least George didn't treat it as a joke and had explained that we were students from Penn State. I have to say my ego was bruised by the fact that my dashing young field biologist look resembled that of a reform school inmate. The quaking bog was astonishing. When you walk in such a place, the "ground," trees and all, moves up and down like a weed-covered waterbed.

Another occasion inadvertently proved that a widely held mythical belief about dragonflies – that they sting – could have been based in fact. In fact, one of our largest dragonflies, the green and blue *Anax junius,* is called the "devil's darning needle" for its association with the devil in folklore and its supposedly vicious stinger. Several of us were out in early May hunting for another big spring dragonfly, *Basiaeschna janata*. This is a gorgeous chocolate brown insect with yellow stripes on its thorax and vibrant robin's egg blue spots along its abdomen. It frequents streams where the females lay eggs in the stems of water plants growing along the edge of the stream. The female crawls tail first down the stem until she is mostly underwater and then inserts an ovipositor resembling a large syringe tip into the plant to deposit an egg. One mistook the bare leg of one of the party for a plant stem and, unfelt, worked her way down to the right water depth to plunge in that ovipositor. He screamed. The dragonfly disengaged and flew off in disappointment. (As I was wearing boots, I could be a fearless and disinterested observer.) The event shows how insect myths might arise from rare errors on the part of the insect. I've handled a lot of live dragonflies. They don't even pretend to sting. Yet under the right circumstances they can make a lasting impression. Hal is now a dignified professor of biochemistry at the University of Delaware, but he maintains his secret life as an entomologist. My sophomore year I shifted to the biochemistry program, which was vastly more to my taste than chemistry. I discovered that a steady girlfriend revolutionized my social life and, improbable as it sounds, greatly improved my grades.

With my rebirth as a biochemistry major (then called ag biochem because we were part of the School of Agriculture), I was able to take a wide range of engaging courses, including zoology, botany, and bacteriology, along with organic chemistry and a course in ancient history that immersed me in the origins of civilization, the blood-drenched joys of Assyrian warfare, the histories of Herodotus, and the plays of Sophocles. Zoology I had some familiarity with, but botany was something new and enormously revealing to someone who had no clue how plants were built or how they worked or were classified. So the "fruits and nuts" lab was a marvel. Microbiology opened up an even stranger world, that of the invisible and dangerous end of the spectrum of life. The lab part of microbiology allowed us to grow living bacteria on semisolid agar at the bottom of a Petri dish. Most were harmless, but in the spirit of the gloriously lax safety regulations of the times, we grew some potentially dangerous pathogens as well, such as glistening golden colonies of *Staphylococcus aureus* (strains of which kill thousands of people each year).

We learned to identify bacteria from the form and color of their colonies grown on agar and by high magnification microscope examination of specimens of the actual bacterial cells that we stained with diagnostic dyes. Individual bacteria are invisible to the eye, but millions bulk up to make a visible colony. In modern microbiology, these colonies are called "biofilms," because they are functional associations of bacterial cells, little worlds where bacteria speak to one another with chemical words. Those discoveries would only come with decades of genetics and molecular biology that we could not even imagine as we gazed at our bacterial gardens. Living bacterial colonies can have quite dramatic colors. *Serratia marcescens,* for example, produces bright red colonies and is likely responsible for a whole genre of medieval miracles involving the spontaneous appearance of spots of blood, like the stories of bleeding communion hosts in damp churches. The age of wonders has not passed. It is possible to enjoy the same miracle with old food items left in your refrigerator. We were assigned to culture bacteria from swabs from our mouths or from a squirt of juice from commercial hamburger meat. We easily grew appallingly large numbers and species of microbes this way. Still, I don't know of anyone who gave up either kissing or eating meat as a result of the experience.

Organic chemistry was my other academic pleasure. The basic concepts of how molecular structures interacted in three dimensions to produce chemical reactions were elegant. I had good hands, so the labs were a joy of doing the reactions and identifying the products. A not-so-illuminating biochemical moment came in my senior year at Penn State. I actually had to take a required class called "Methods of Agricultural Analysis" taught by Professor Herald O. Trebold, the soon-to-retire head of the Biochemistry Department. Sound boring? The course seemed like a throwback to a sort of cornfield dawn of biochemistry, a parochial ag school anachronism, perhaps like requiring medical students to take a class on preparing willow bark poultices and learning how to apply leeches for therapeutic bleeding. I well remember one session in which we learned how to properly scoop wheat from the sample ports of a railroad boxcar. Sadly, we didn't actually have a loaded boxcar on hand to clamber over. This was a nicely diagrammed theoretical lecture about boxcars. But the class did have a wet lab, which consisted of innocuous activities like determining the nitrogen content of a pound of grain from a scoopful that might have once been in a boxcar.

As for evolution, my classes didn't include much mention of it. I suspect that my concentration in biochemistry assured that I would be exposed to a kind of biology that accepted evolution and then didn't trouble its head about it any further. As a consequence, I would first hear about the creationist outlook on the world at Penn State as the target of a self-nominated missionary. I had gotten to lunch late and was sitting alone at a table in the cafeteria. A student whom I didn't know put his tray down on the other side of the table, saying, "Do you mind if I sit here?" I said no. He sat down. He arranged his silverware and blandly asked, "Have you accepted Jesus as your personal savior?" I owned that it wasn't one of my objectives. He was unperturbed and continued to tell me that his faith included belief in the absolute truth of the Bible and of the creation of all life as we see it today. I asked him his major. "Pre-med," he replied. "Don't you have to take biology?" He said that he did. I raised what I hoped was an awkward possibility. "What do you do if they discuss evolution?" He waved that off. "I just learn it and give it back on the exam. I don't have to believe it." Here was a mind self-confined in a chamber of brilliant unalterable certainty. Evidence and the natural world meant nothing in the face of faith. I innocently thought that he was an isolated oddity.

I spent much of the summer between my junior and senior years literally at sea in the Cold War. This was the much-anticipated six-week summer training cruise that all midshipmen took. I was assigned to the oceangoing minesweeper USS *Salute* based in Charleston, South Carolina. I didn't realize it at the time, but I was one in a fading line of U.S. Navy sailors who would man a wooden warship. A glorious tradition in naval warfare going back thousands of years to the ancient sail and oar-driven rams, the Greek triremes, still existed in the form of seagoing minesweepers. The ship was small, noisy, and packed with machinery and the constant smell of exhaust. Our quarters were hot and crowded but well ordered. The *Salute* had a 35-foot beam, a 14-foot draft, and a length of 172 feet, and this round-hulled wooden ship rolled as handsomely as a rubber duck in a roiling kiddie pool. Even in light seas one could stand on the deck and watch the horizon vanish and slowly reappear as the ship rolled on its long axis. The permanent crew consisted of six officers, four chief petty officers, and 67 enlisted men. For historical comparison, Darwin's ship, the *Beagle,* had roughly the same size crew and was only ninety feet long, and Darwin's voyage lasted five years. I'm sure that our food was better; we had excellent navy cooks, refrigeration, and fresh meat, and we lacked weevils in our flour. We were never out at sea for more than a few days at a stretch.

Duty on such a ship is generally slow and tedious. Most days we were tied up motionless and breezeless to a navy pier in Charleston Harbor, with the Sun ever so slowly crossing the hot, hazy sky. It reminds me now of what Mark Twain once said about shipboard life: "I used to like the sea, but I was young then, and could easily get excited over any kind of monotony." It wasn't always dull, though. The first time I was within touching distance of wild dolphins was from the bow of the ship's whaleboat while we were out inspecting mine-marker buoys. The dolphins were clearly having fun and giving us the eye – literally from eye level as they surfed our bow wave and deliberately goggled us as strange specimens from another world.

Soon I went back to Penn State for my senior year. This was during one of the tensest and most dangerous of Cold War periods, the October 1962 Cuban Missile Crisis. The Soviet Union was discovered to be installing nuclear missiles in Cuba. The United States declared a "quarantine" of ships carrying weapons to Cuba, which the Soviet government regarded as a blockade, potentially an act of war. We senior midshipmen were informed in class one morning that we would likely not graduate but would be commissioned and within days be assigned to duty stations. The looming possibility of a nuclear war did not resonate well as a viable career move. My wife tells me that her roommate at the time came home in tears upon hearing the same news from her boyfriend. No worries. She soon got another boyfriend.

I recall watching the grainy news footage of missiles being transported to Cuba on Russian ships and listening to the escalating rhetoric on the news with a feeling of futility as the thing played out. The U.S. Navy intercepted Soviet ships. The crisis ratcheted upward, awaiting the critical confrontation that would set off a nuclear weapon and the inevitable response. Fortunately, before some final, irrevocable incident, cooler heads averted nuclear war by opening direct negotiations between Kennedy and Khrushchev. I think none of us really comprehended how precarious our world was and how casually it might have ended. The ordinary lived on, and no door opened into a post-nuclear dystopia. Yet the ordinary that survived those days was a new ordinary. This is what we later enjoyed as the Strangelovian military dogma of "mutual assured destruction," which reassuringly oversaw decades of the Cold War. The Soviet Union collapsed twenty years ago, but bad habits die slowly. We still senselessly aim nuclear warheads at Russia, and they return the favor.

I took my first course in invertebrate zoology from Alice Beatty during the summer term after I had graduated. My parents were moving to Pullman, Washington, and I stayed on to work for a summer at Penn State. Having grown up in Pittsburgh, a thousand miles from the sea, I had never imagined the existence of such animals as those elegant ocean dwelling invertebrates revealed by Alice's lectures and lab. On land the macroscopic invertebrates are mostly insects with a smattering of earthworms, millipedes, pill bugs, spiders, ticks, slugs, and snails. In reality the vast majority of the noticeable animals on land belong to just two great clades that had evolved on land once plants had gained a firm foothold, the insects with perhaps several million species and the tetrapod vertebrates (salamanders, frogs, reptiles, mammals, and birds) with their paltry 12,500 species. These are the obvious ones. A third lineage, the unobvious and most prolific phylum of land-dwelling animals, is made up of the roundworms, the nematodes. They outnumber all the other animal species put together.

I say that because although there are many free-living, nonparasitic kinds of nematodes, the majority of members of this frightening hidden kingdom live in a well-protected environment, mainly, the insides of the rest of us. They are internal parasites of the other land-living creatures, including those well known to dog owners: hookworms, intestinal roundworms, and heartworms. As each species of insect likely harbors its own species of nematode parasite, there should be at least as many nematode species in the world as there are insect species – in the millions. Most nematodes are tiny worms, but just for cocktail small talk, it is helpful to know that there are some substantially bigger ones. The largest nematode is called *Placentonema gigantissima*. Its niche is as a parasite that lives in the placentas of sperm whales, and it comes in at a hefty thirty feet in length. That's longer than all but the largest recorded pythons.

Most of the thirty-five phyla of animals live only in the sea; only a handful of phyla were ever able to evolve adaptations to endure the rigors of life on land. All the strange, otherworldly marine invertebrates – flower-like anemones, rippling marine worms, charming hermit crabs, wide-eyed squid, jellyfish, sea pens, brachiopods, bryozoans, sea urchins, rotifers, horseshoe crabs, sea cucumbers, sea squirts, sea spiders, starfish, and many others I first met in my invertebrate course – hooked me, as did a short field trip I took after the class with George and Alice to Woods Hole and Nantucket. This was my first ever experience snorkeling in the sea, searching for marine invertebrates in their domain. The ocean lives after sunset too. We went with masks and flippers into the night sea off the Marine Biological Laboratory boat dock (the laboratory is affectionately known as the MBL to generations of biologists). The late summer water was full of plum-sized, nearly transparent jellyfish-like creatures called ctenophores, which at this time of year seem to hold a gelatinous annual convention at Woods Hole. When disturbed by our swimming among them, hundreds of ctenophores' ciliary comb rows pulsed with brilliant colored light in the black water. I also became an aficionado of spear fishing for flounder in an inlet on Nantucket Island. It's all a matter of learning to spot the funny pair of eyes sticking up out of the thin layer of sand under which they conceal themselves. If they see you coming, they sensibly dart away, but then settle down again in plain sight – all threat forgotten. If you can hold your breath long enough and can aim a spear well, they become satisfactory fried guests at dinner.

George and Alice introduced me to the concept that the body plans of animals had to have had an evolutionary origin. There had to have been common ancestors to the distinctive animal phyla of our present world in the deep past, and some zoologists had thought deeply and fantastically about what they might have been like. Around the start of the twentieth century this enterprise flowered into a whole industry of trying to derive vertebrates from ancestors in various other phyla – including from horseshoe crabs by front to back inversion of internal body parts. Our hearts are in front and our nerve chord in back. It is the reverse for arthropods such as horseshoe crabs, lobsters, insects, and many other invertebrates. To me, this use of comparative anatomy meant we could know about the past through biology as well as from fossils. I first heard about Gavin de Beer's *Embryos and Ancestors* at this time, from George, who constantly threw out the titles of interesting books. I bought a copy and would finally read it during my first semester in graduate school at Duke. This helped occupy otherwise boring Saturday morning chore sessions at the nearby White Star Laundromat. Sometimes I had a soft drink and sat in the Little Pig Barbeque just across the parking lot for a change of scene. De Beer outlined his ideas for me about how evolutionary changes in embryonic development could have led to large changes in animal evolution. Although I really didn't have the background to understand completely the arguments and examples derived from embryo and larval forms, it was an amazing eye-opener and served as my entry point into what I and other colleagues would eventually develop into a new field, evolutionary developmental biology, or evo-devo.

But first I would have to leave off being a scientific larval form myself.

Nothing is inevitable until it happens.

A. J. P. Taylor

SEVEN

Going South

BIOCHEMISTRY

Near the end of my undergraduate life, I set about blissfully applying to graduate schools, including Duke. There was lurking a possible slight hitch to entering this dream world; I had a commitment to serve two years of active duty in the U.S. Navy after commissioning as an ensign (the naval term for what the army calls a second lieutenant). However, the navy seemed to have had an excess of young officers entering at the time and readily allowed anyone qualified to have eighteen months in the inactive reserves to get a master's degree before going on active duty. The master's degree limit seemed like a constraint, but I hadn't met Bill Byrne yet. Bill was a biochemistry professor at Duke who served as the departmental graduate program director. He was undaunted by both the department's and the navy's rules and worked out an acceptance for me to take a master's degree, which the navy approved. He then produced a second letter to convince the navy that as Duke didn't really like to give a master's in biochemistry, it would be far better to let me stay on for a Ph.D. In a miracle of bureaucratic accommodation, they approved that too.

I started graduate school in August 1963. My first impression of North Carolina was of heat, humidity, and the penetrating 4:00 AM singing of mocking birds just outside my window. The excitement of seeing a species of bird I hadn't seen before couldn't mollify that. The sweat of trying to find my way around was compensated for by the rich diversity of trees on campus. There were some familiar northern ones, such as the enormous sycamores, and lots of southern species, such as laurel oaks, magnolias, and camellias with their glossy green leaves. Camellias strangely produced their large tropical blooms in the fall. I would later discover spots in the eastern lowlands of North Carolina where damp bald cypress trees were festooned in Spanish moss (in truth neither Spanish nor a moss but a flowering plant epiphyte related to bromeliads – a family that includes, among other things, pineapples).

I can't say I enjoyed my first semester courses at Duke (at least the ones over in the Chemistry Department, dreary and horribly taught by somnolent instructors), but I vastly more enjoyed my spring biochem classes. Given my usual non-predilection for math, my favorites unexpectedly included the physical biochemistry of proteins course. That was taught in part by Professor Charles Tanford, one of the leading American physical biochemists, who was to my surprise (based on my initial dread of his topic) an excellent teacher. His lectures revealed macromolecules as measurable physical entities. Tanford could be funny, especially when he wasn't meaning to be. The best Tanfordism I ever heard was at a party, when admittedly the wine was freely flowing. We were talking about the human population explosion, and Tanford in his grave manner pronounced his solution: "If we could teach everyone in the world the equation for population growth, they could draw the growth curve, they would see the implications for themselves, and of course limit their family sizes." A novel approach certainly, but it seemed like a far-fetched proposition even for an academic.

The co-instructor of the proteins course was Professor Robert Hill, one of the pioneers in the evolutionary study of protein sequences – through the evolution of molecules themselves. Emil Zuckerkandl and Linus Pauling, in a seminal paper written in 1962, laid out the foundations of molecular evolution, which pointed out that genes could evolve as "molecular clocks" and that protein sequences (and later, when the technology was developed, gene sequences) could be used to infer the phylogenies (evolutionary relationships) of organisms. New methods that let biochemists to take apart proteins in a stepwise way to see how their chains of amino acids were linked were allowing the dissection of these molecules in order to understand how they worked and evolved. The realization that the sequences of amino acids in the

My parents' wedding, Shawinigan, Quebec, 1939.

(*Above*) My father with his parents, Vienna, ca 1935.

Intimations of two vastly different temperaments: (*facing*) my mother and the circus elephant, ca. 1934; (*left*) my father reading on the porch, Lac Souris, Quebec, 1955.

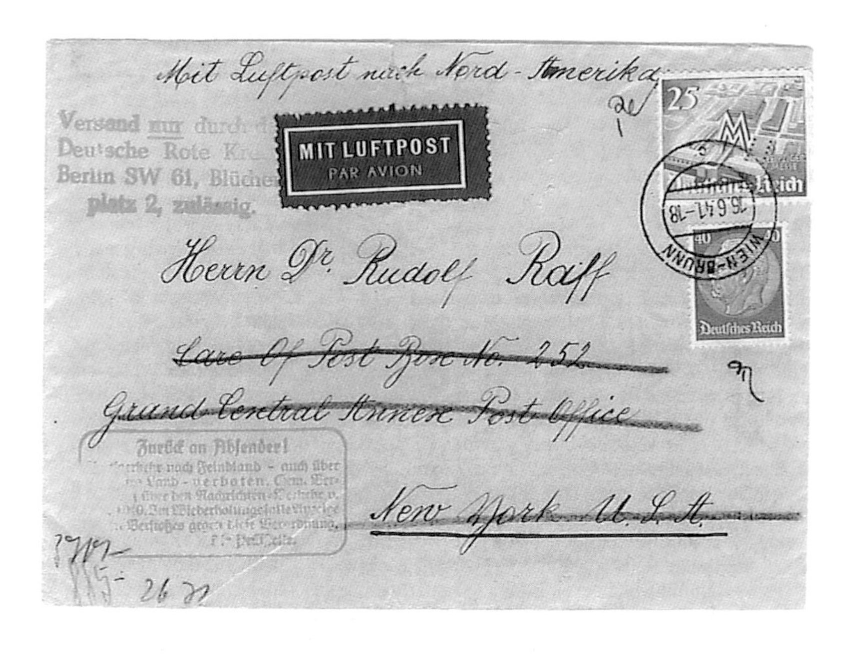

An envelope sent to my father by my grandmother in 1941; intercepted by the German censor and returned to her with a warning.

Photographs courtesy of Ed Fraser.

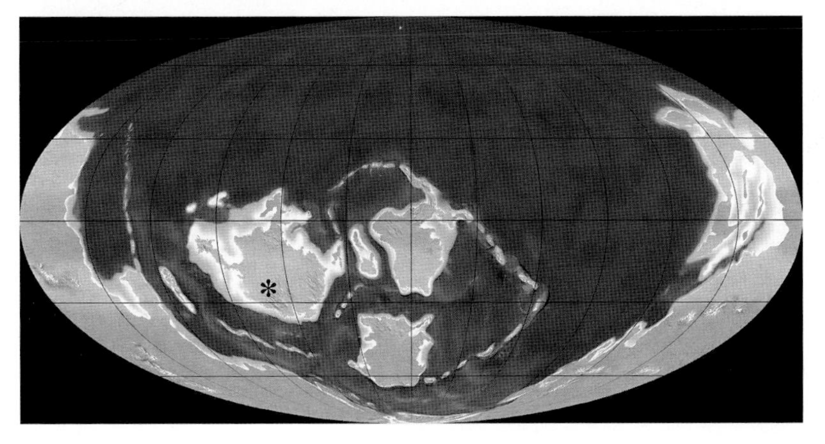

(*Top*) Driving my grandfather's motor boat, Lac Souris, Quebec, 1952. (*Lower*) Half a billion years ago, during the Cambrian explosion of animal life, the ancient continent of Laurentia lay just south of the equator. The approximate area of the paleocontinent where present-day Quebec will be located is marked by the asterisk.

Paleogeographic map (Mollweide globe) of the Earth used by permission, courtesy of Ron Blakey, Colorado Plateau Geosystems.

Fieldwork in Chiapas, Mexico, 1969. (*above*) George and Alice Beatty cataloging dragonfly collections at day's end. (*left*) A Maya stele in the square of a Chiapas village. The local people still spoke Mayan.

(*Above*) Beth by a sun-baked afternoon streamside.
(*left*) Me in the shade.

Photographs, Beth Raff and Rudy Raff.

Keeping company with a friendly feral pig, Chiapas, 1969.

Photograph, Beth Raff.

same protein from different creatures could reveal their degree of evolutionary relationship was an extraordinary insight and eventually led to the great confirmation of Darwin's ideas on relationships of organisms to each other by a completely new kind of data. Perhaps I should have joined Hill's lab studying the evolution of hemoglobin genes, but I was disinclined by the regimented way in which protein sequence labs were run. Each student did a dissertation based on the tedious multiyear labors they had invested to determine the amino acid sequence of a single protein from a single species. The boss then did the fun part of synthesizing the evolutionary relationships from all the students' results. One protein sequence was years of dedicated work. Now with gene sequence techniques you can determine the sequence of a whole bacterial genome in days.

I joined the lab of Robert Wheat, a microbiologist interested in the biochemistry of bacterial cell walls. Bacteria are surrounded by envelopes that essentially consist of a single giant molecule built by the linking together of molecular subunits secreted by the bacterial cell. They are important as a site of antibiotic action and in the recognition of bacterial infection by the immune system. I studied one of the molecules of a bacterial surface. The biochemistry I did was enjoyable and earned me my degree, but I came to realize that I was really doing a fascinating organic chemistry problem that did not draw me in intellectually and did not advance a wider understanding by much. I think I was sufficiently immature in science when I chose my lab that I could not see that one should think of how to ask the big questions and how to devise feasible experiments to help answer them. I was still puzzle solving at a level removed from a real understanding of what makes science. My mental engagement was elsewhere.

Most of my reading was on organismal and evolutionary topics. There were some compelling papers in which molecular-level approaches were powerfully revealing answers about evolution. I especially was struck by the work of P. P. Cohen, a biochemist at the University of Wisconsin who showed that the evolutionary shift from secretion of ammonia by fish to secretion of urea by land animals was recapitulated in development from tadpole to frog. I could see that molecular biology was transforming biology, but at this point, I wasn't sure what kind of biology I should go on to in order to build a research career. I felt less and less enticed by the biochemistry that had attracted me to Duke. The hint I should have heeded was that I far more enjoyed seminars in the Zoology Department than in my own department. I was a faithful attendee of zoology talks. I also learned some interesting lessons about natural history from zoology graduate students, like the one I gained from a friend, a graduate student we called "John the spider man." He saw more arachnid life than I had ever thought of looking for. A trip to the woods with John meant seeing a world spilling over with spiders in every flower and under every leaf. I've never been that fond of spiders, but I confess I did find a liking for the little jumping spiders. They appear to engage visually, and so seem conscious.

In January 1964, during Christmas break in my first year at Duke, I met Beth Craft. I was going to spend part of the coming summer with my parents in Washington, but I couldn't afford to fly to the Pacific Coast for Christmas holidays as well. Instead, I drove up to Penn State for a few days to visit George and Alice Beatty and Hal White. Before I left, Bill Byrne – ever the eager recruiter of graduate students – suggested that I advertise the Duke biochemistry program to any potential students I might run into on my visit. I mentioned that to my old undergrad biochemistry professor, Carl Clagett, when I stopped in to see him. He said (more prophetically than he could ever have imagined), "Oh, I'll introduce you to Beth. She'll take care of you." Beth was working in the lab across from his office. Scientists like to think that when we enter a lab, fortune will favor the prepared mind. I wasn't prepared, but sometimes fortune simply favors the very lucky. We had only a few days to see each other then, but we corresponded over the next few months – an indecisive way to establish a relationship. I went back to Penn State that summer so we could talk about the future. We drove out to Ten Acre Pond, a sunlit water lily pool shimmering with dragonflies. This big blue sheet of water lay in a large forest clearing, the naturalized remnant of what had been an early-nineteenth-century bog iron mine. It was a quiet place to sit together. Beth had sufficient credits to graduate at the end of her junior year. She applied to Duke and was admitted to the fall biochemistry graduate class.

When we announced our engagement to our parents, both our fathers (in a charming old-fashioned way) looked each other up. They seem to have been satisfied with their researches, and they discovered a coincidence. My father, and the rest of the family, entered the United States under a post–World War II program intended to bring in talented European scientists. A number of these scientists naturally ended up in government projects. The catch with the highest visibility was the rocket engineer and space visionary Wernher von Braun, who had designed the German V-2 ballistic missile. He would take the U.S. space program to the Moon. My father brought no rocket technology. He was a polymer chemist and thus relevant to the coming plastics revolution. His admission to the United States was sponsored by Koppers Company, which had an expanding research program centered in Pittsburgh. Each foreign scientist recruited postwar was enabled to immigrate into the United States by an individual act of Congress that was titled something like "An Act for the Relief of Dr. Fill-in-the-Blank." Beth's father, Ed Craft, who was the legislative counsel to the U.S. House of Representatives, realized that he, in his younger days in the office, had been the one who drafted the bill that brought our family into the United States.

Beth and I were married on June 12, 1965. Our honeymoon was an elaborately choreographed camping trip in the American West that ended at my parent's house in Pullman, Washington. This may seem like a peculiar honeymoon, but only my mother had been able to come to the wedding, so only she had met the bride. Beth had seen the Grand Canyon and other western sites on a family trip when she was a teenager. I had once been to Washington and eastern Oregon to visit my parents, but never into the romantic Rocky Mountain West or the northern Plains. We traveled by car and camped under canvas most of the way. We hadn't yet heard of lightweight nylon tents, so we bought a remarkably bulky, heavy, and complex tent complete with a suite of poles, stakes, flaps and ropes that when assembled looked like, and was nearly large enough to have served as, Julius Caesar's Gallic War campaign headquarters. We loaded that canvas monster into our car, the notorious "unsafe at any speed" Corvair. The space remaining was filled with various camping gear, stove, pots and pans, sleeping bags, clothing, and much else we somehow were able to fit in. After the wedding we followed the trail westward.

My father, whose vision of camping came from a trip he had taken to Turkey in the late 1920s, had no idea of what camping in national parks entailed or permitted and told us we should take a gun. My mother helpfully sent us every newspaper clipping she could find about fatal grizzly bear attacks on women in sleeping bags. For years she assiduously kept us up to date on all bear mauling stories anywhere in the forty-eight contiguous states. Beth says she still remembers the details of each dreadful story provided by her mother-in-law. We never had problems with bears when hiking or camping deep in the bush, but one night in a national park, we found them disquietingly overly familiar. They roamed freely through the car camping area looking for handouts from people who foolishly obliged and thus acclimated the local bears to associate humans with food. We piled all our gear into the tent and slept in the car.

Our trip led us to all the standard big western geological and natural history tourist stops: Devil's Tower, the Black Hills, the Badlands of South Dakota, Yellowstone, Glacier National Park, and the preposterous although imposing Mount Rushmore. The hiking in such places was a revelation. Sun, rocks, glaciers, geysers – all the grandeur that had enthralled the nineteenth-century painters and naturalists of the days of the great West – was there to see and touch. For someone who had never been in the West before, the effect was one of being overwhelmed by the brilliance and clarity of light in the dry air and the vastness of the scenery. The outsized scale of places like Devil's Tower distorts perception. The tower is an ancient volcano neck left behind from a much larger volcano cone, which has long since eroded away. The resistant hard-rock neck is roughly a thousand feet tall, a flat-topped pillar composed of vertical hexagonal basalt ribs. We climbed up to the base and circumnavigated the tower amid the vanilla smell of ponderosa pine baking in the intense light and heat of the midday sun. Up close, the thin ribs became monstrous basalt hexagons, thicker than I was tall. The hexagons were the product of cracking that took place as the molten basalt slowly cooled and solidified inside the ancient volcano.

The café where we had breakfast had pine-paneled walls decorated with elegantly varnished wooden plaques, each bearing a "jackalope" head. Jackalopes seemed to be the state animal of Wyoming. Unlike most state mascots, jackalopes are made, not born. They start out as stuffed jackrabbits – sometimes entire rabbits in life-like pose but generally just the heads mounted on plaques. A jackalope enters the world when a taxidermist convincingly attaches the forked ends of a pair of pronghorn antelope horns to the rabbit's head. These chimeras are surprisingly appealing to some – and appalling to their wives. I admired them greatly and have wished ever since that we had bought one. The monument lacked living jackalopes but compensated with colonies of black-tailed prairie dogs.

Although we were the greenest of novices in our first big camping expedition, camping and especially canoeing became a major part of our lives for many years. We would two years later make the thirty-six-mile round-trip hike along the Hoh River through the Hoh rain forest to its source at the toe of the Hoh Glacier in Olympic National Park. The hike took us from sea level to an altitude of seven thousand feet through the hemlock and spruce rain forest into the sub-alpine zone, where trees thinned out to mountain meadows – all covered with snow when we arrived. We'd have to wait until another time to reach the flowering summer alpine meadows. The wet, heavy snow blocked our slog the last few hundred feet to the glacier. The surface was undermined by invisible but dangerous melt channels that we could hear gurgling beneath the snow. We met one other couple when we got to our campsite below the glacier. They hadn't limited themselves to easy-to-carry food but had collected clams on the Pacific Coast in the morning and carried them in a bucket of sloshing seawater the full eighteen miles to have them for dinner. Impressive – and frankly nuts. On the way down we slept in the grassy flats of Elk Meadow. No elk in sight, but plenty of encouraging elk droppings covered the ground. The towering cliff walls made night come to the valley floor while the sky was still blue above. That blue deepened into black soon enough, leaving for our only light the brilliant band of stars of the Milky Way. We suspended our food as high in the branches of a convenient tree as we could manage to rope the sacks. No one raided our granola during the hours of darkness. The Olympics would later become frequent hiking grounds for us once my parents retired to the Olympic Peninsula.

We bought a canoe in 1966 with a surprise bonus from the Biochemistry Department stemming from a discrepancy between stipend promised and actually paid over our first two years as students. We drove up to Washington, D.C., bought a Grumman aluminum canoe, paddles, ropes, and life preservers, tied the canoe to the roof of the car, and went to the Capitol to have lunch with Beth's father to show it all off. Grumman canoes were made by the aircraft company known in World War II as the Grumman Iron Works for the sturdiness of their warplanes. That canoe weighed seventy pounds. Neither of us had actually ever been in a canoe before we bought the thing, so we had to deal with the awkwardness of learning how to operate it. At first we zigzagged across a little lake near Durham and argued about who was paddling wrong. As I was sitting in the stern, the steering position, there was no disguising that I was the only one responsible for our creative and unpredictable track through the water. I finally did get the J-stroke down.

We used that canoe for the next forty years, including on two long trips, the first at Algonquin Park in Canada and the second in the Boundary Waters, the canoe route eighteenth-century voyageur fur traders traveled. There we had to pack in food and supplies for two weeks. In both trips we were able to paddle and portage miles into the backcountry with only map and compass. This was an unparalleled way to see wilderness – if you had intact mosquito netting on your tent. Otherwise it was intolerable, and we met people with eyes nearly swollen shut who had turned back after spending a night without netting. We were serenaded by the mad song of a pair of loons, lay on our backs on the warm granite of a small island, and watched the northern lights, caught fish, had a beaver visit our camp. We spent one night huddling in a thunderstorm on a small, rocky island where we had camped. We had set up our tent under the few trees on the island, but we soon thought it better to hunker down on the less comfortable rocks at the bare end of the island than to sit in our cozy lightening-target zone under a tall tree.

Eventually our hiking and camping habits shifted from the wet and lush North to the alien deserts of the Southwest. We walked into the ruins of thousand-year-old cultures, clustered into hard-to-reach rock shelters perched on thousand-foot cliffs. We learned from walking into the delicate pink- and violet-banded draws of the dinosaur-bearing Morrison Formation that Georgia O'Keeffe painted the New Mexico landscape as she actually saw it. We camped at Scorpion Corral, and we were immolated by light on the white, wind-rippled gypsum desert of White Sands.

One after another, members of our northern family would be drawn to Duke, a genial southern university, like fruit flies to a glass of red wine on a summer day. Beth's sister Alice turned up as an undergraduate in 1964, the same year Beth started at Duke. Alice, who dreamed of leaving home and finding independence, was at first annoyed by her sister's choice but soon discovered that having relatives in town was useful as a device for getting away from the dorm for the night. Just after we left, my sister Mimi started at Duke in physiology, where she met her future husband, another graduate student, Laszlo Jakoi. As both Beth and Mimi were identified in Duke records as E. Raff, Mimi after a couple of years started getting notices from the graduate school that her time spent working at a degree was getting too long and she would lose her eligibility. The confusion was finally sorted out.

Finally, in the 1990s my brother Bob and his then-wife were getting divorced. Under that familial Duke-ward compulsion, Bob quit his technician's post in the organ transplant research unit at the Fred Huchinson Cancer Research Center in Seattle to register for the Duke University physician's assistant program. As one of his required clinical rotations, he worked for a few months in Maine. There he met Mary Delois, a nurse practitioner. They would settle in her hometown of Portland, Maine. Bob thought that Portland was pretty similar to Seattle if one didn't pay too much attention to the four-foot mounds of snow and the hard blue ice of a long Maine winter. They were married in 1995 on a crisp fall day. My mother loved their wedding. She was, however, someone who sometimes let her enthusiasm overwhelm her discretion and was notorious for singing into the soloist's part in church services. As we settled into the pews of the rural Maine church they had chosen for the ceremony, my mother sat between Mary's sister and Amanda, our daughter. My mother turned to Mary's sister and opined brightly, "I'm so glad Bob is marrying Mary. I never liked those other women." Amanda blushed, but no else one seemed to notice.

RACE

My first years at Duke taught me about another part of American life that I had been amazingly ignorant of as I grew up – race. For someone who had been raised in America that seems laughably unlikely, being so oblivious, unable to glimpse the proverbial elephant that filled the room. However, I had been raised in the sheltered suburbs of a northern city. Northern cities had black residents, but the suburban neighborhoods were largely white. There were only a few black children in my schools, few even at Penn State, and none in Navy ROTC in my time. Our schools did not discuss race. When I drove to Duke in 1962 for my graduate school interview, my father took a couple of days off to come with me as a driving companion. We stopped at the Natural Bridge in Virginia, a spectacular towering natural stone arch spanning the bed of the modest stream that over time had cut the arch. This kind of erosion structure is rare in the eastern United States, but the arch wasn't to be my greatest surprise. I was shocked to find when I went into the visitor center that there were categories of plumbing I'd never seen before. There were "Whites Only" and "Blacks Only" bathrooms and water fountains. Yes, they had obvious segregation in the South – even at a (privately operated) national landmark. I would learn that we had segregation in the North, too, but in a more subtle way, achieved by separating where people lived. Whites and blacks actually lived in much closer proximity and interacted more in the South, and once I enrolled at Duke, I would meet many more black people there than I had in Pennsylvania.

Segregation was conducted somewhat differently in the South, where it had a stronger ideological basis than in the North. In the North, neighborhoods were segregated not by overt law but by banks, which made it close to impossible for black families to arrange mortgages in "white" neighborhoods. This method gave segregation an economic foundation with deniability that race was involved. Economic segregation in the North was of course not the only means of enforcement of inequality. Urban black communities felt occupied by white police. Economic inequality and incidents of abuse by police led to the rioting of the black neighborhoods in northern cities in the 1960s. White northerners were surprised, showing how little real contact there was. The concept that prevailed with respect to schools was the convenient fiction of "separate but equal," based on racially segregated neighborhoods. No wonder I so rarely saw a black student in my classes. School segregation produced inequality of education and opportunity and was declared illegal in 1954 through the force of a U.S. Supreme Court ruling in *Brown v. Board of Education of Topeka.* The first case decided against de facto segregation in a northern school took place in New Rochelle, New York, in 1961, but it's hard to say that such compelling rulings have equalized education fifty years later.

Progress in America is famous for following a rising sawtooth curve, but the sawtooth curve is in some ways tending down now. According to an editorial by Ruth Marcus writing in 2006 for the *Washington Post,* "Public schools are less integrated today than they were in 1970. In the South, many school systems, once segregated by law, have been freed from court oversight and, with the return to neighborhood schools, have reverted to their former state." And as for the economic segregation that was present in the North when I first became conscious of race, it's still going on. In his book *Searching for Whitopia,* Rich Benjamin, who himself is black, has written a lighthearted but potent story of current race relations by seeking out and visiting with the white participants in the flight from diversity. There he goes on to befriend, play golf with, and talk at length with people who have moved to the improbable new remote enclaves of white flight. The people he talks with are refugees from racially mixed America, for example, migrating from Los Angeles to fast-growing white-populated developments in Saint George, Utah, or Coeur d'Alene, Idaho. These are not just stories of weirdness in America. This is only the latest episode in the de facto segregation inherent in the decades-long movement of the predominantly white American middle class to the suburbs. Economic activity has followed, and so have the lion's share of social services and support for schools.

The more personally shocking racial interactions I saw were in the South. Because I had grown up a northern suburban kid, I had no experience of southern culture and of what rules regulated its high level of direct daily contact between blacks and whites. This would be the first time I would be able to feel even briefly and superficially what being a member of a minority meant. We were selling our car. A student from a black college in Durham was interested in looking at the car and called me up. He had no transportation, so I drove over to his campus. I parked and walked across campus to the building where he would meet me. I was the only white person there. Everyone was pleasant, but I realized that where black was the norm, white stood out. I was the other, the obvious one, the stranger, the outsider. For a brief time, I could stand outside of myself and see the inversion. I had the sense of what it must be like to have to live daily as the vulnerable member of a visible minority, but I had never had to live it.

There was more day-by-day interaction between the races in the South than in the North, but as in the North, the force of a pervasive economic segregation was present too. The poor side of Durham was the black side of town. Poverty was combined with a history of explicit expression of racial discrimination and control of black people by horrendously violent suppression. In 1960, a major step in integration took place in Greensboro, North Carolina, just fifty miles from Durham. Four black students sat at the Woolworth lunch counter and requested service. They were asked to leave, but refused. That brave demonstration was the beginning of a series of major lunch counter sit-ins, which eventually involved thousands of black students and were a major catalyst in the civil rights movement. I knew so little about them as a northern newcomer that I didn't even realize that these events had begun near Durham and had occurred only three years earlier.

Some racial incidents from my first days in the South still stand out starkly for me. The worst was the bombing of a black Baptist Church Sunday school in Birmingham, Alabama, in September 1963. I had only been in Durham for a month when one morning I turned on my car radio. The news was of the deliberate killing of four children in that attack. It was a heart-stopping brutal act. I pulled to the curb and sat stunned; I had to suppress tears. The callous attitude that led to the bombing of children was openly and cheerfully celebrated by some exhibitors at the North Carolina State Fair that Beth and I attended two years later.

Beth had always liked going to the Indiana State Fair, and so she wanted to go to see how it compared to the fair in North Carolina. It was hot and crowded, and it had the standard livestock barns and prize-winning oversized vegetables, but this state fair was just a little different in another way. Among the participating organizations, with fair-sanctioned exhibition booths, was the Ku Klux Klan. The Klan's booth included a loudspeaker that played racist songs that could be heard across much of the fairgrounds and a crew of thuggish characters ready to assault anyone who complained. Their outrageous behavior offended enough fairgoers that they were banned from the state fair in subsequent years. The booth next to theirs was more subdued but equally grotesque. It was operated by the John Birch Society and had at least one off-duty Alabama deputy behind the counter. The John Birch Society was in those days at the leading edge of right-wing paranoid politics in America. The society backed the notorious segregationist police commissioner Bull Connor, who dominated the Birmingham police force at that time. His single-minded approach of suppressing peaceful civil rights demonstrations with dogs and fire hoses led to the rise of Martin Luther King.

Ultimately the events in the Alabama cities of Montgomery and Birmingham, and particularly the Selma-to-Montgomery march, produced national revulsion with segregationist violence. King's demonstrations paralyzed Birmingham, and Connor lost his bid for mayor. Beth was openly skeptical of something the Alabama cop at the Birch booth said to her. He became red faced and told her he was from Birmingham and she was lucky she was a woman. Perhaps, unlike the Quakers who advocated "speaking truth to power," he felt that one should speak power to truth.

Durham itself at that time was home to a virulent hate radio commentator named Jesse Helms whose broadcasts we occasionally heard. Astonishingly to us, in 1972, some years after we had graduated from Duke, this divisive broadcaster was elected a Republican U.S. senator. It is a palpable understatement to say that Helms was no statesman, but he proved to be a skillful schemer. He worked his remarkably effective talent as a demagogue and spoiler in the Senate, and shamefully, he had no difficulty being reelected until he retired in 2003. Helms helped build the conservative backlash in American politics by exploiting the American racial divide throughout his Senate career. He steadfastly opposed civil rights legislation, school integration, and voting rights. He was proud of his sixteen-day filibuster against the bill that established the federal Martin Luther King holiday.

The Ku Klux Klan would appear en mass in Durham in 1966 to protest a visit by Hubert Humphrey, then the vice president and in 1968 a Democratic presidential candidate. While Humphrey was speaking to an audience in Durham, KKK members who had dressed in their Sunday best bed sheets gathered together in the muggy North Carolina night, milling around in the corn stubble of some sympathetic farmer's field. There they burned their trademark cross. The previous evening we had eaten in a diner we normally liked and found the booths packed with Klansmen and their uniformed "security guards." Out of curiosity we lingered and listened to the snatches of conversations around us. Most of what they said was banal, but the presence of a roomful of unabashed Klansmen was oppressive and gave us an inkling of the suffocating weight of being surrounded by brownshirts and ignoramuses. Times were changing in important ways, though. In Joe Blum's lab, where Beth did her Ph.D. work, a fellow student, Ida Owens, became the first black woman to receive a Ph.D. from Duke. Yet that kind of triumph has to be weighed in balance with continued racism.

In the years following public school desegregation, southern states found ways to get around the change by founding segregated private academies. Persisting racism was exploited politically by Richard Nixon in his "Southern Strategy," which drew masses of southern conservatives away from the Democratic Party to the Republicans by playing to white discontent over desegregation and a perceived favoritism in favor of black people. Ronald Reagan successfully mined the same discontent by creating the narrative of the "welfare queen" (black naturally) who got rich from the welfare paid for by taxes on white workers. The tactic worked, but over time, Republicans have paid a greater price than they know. Once this was a party that could tolerate nuanced views on issues and could govern; it now has been hollowed from the inside and molded into a chorus of shrills who preach theocracy and racial resentment.

Forty years after the murder of Martin Luther King in 1968, Barack Obama, the son of a black Kenyan father and a white American mother, was elected president. His inauguration was a brilliant and moving moment in U.S. history. Well, not to everyone. Extremist conservatives were not able to accept the legitimacy of a black president and so have denied that he was born in Hawaii, claiming that his record was falsified to cover a Kenyan birth – despite the existence of his Hawaii birth certificate and a birth announcement in a Honolulu paper. That was only the start. Accusations that Obama is a Muslim, a socialist, and a racist hater of white people were soon to follow. An amazing number of white power groups still thrive in America. They display high levels of paranoia and affection for violent language and even outbursts of violent action against black or Hispanic people. These associations mutate, form bitter factions, and then reform in new guises. They and the more "respectable" conservative politicians and media commentators are the heirs of Jesse Helms and soul mates of a gaggle of diehard holdouts huddling together trying to revive the mummified corpse of the Confederacy. If that sounds a bit harsh, consider the governor of Texas's 2009 red-meat suggestion that secession from the United States might again be a good program for Texas. A party of intractable political opposition to President Obama was born to a marriage of white supremacists, ultra-conservatives, and Christian dominionists. Many claim not to be racists but are happy to exploit racial prejudice under other, more acceptable cover names. These politicians easily achieved national reach through a drumbeat of conspiracy-laden propaganda perpetuated by a network of right-wing radio and television talk show hosts.

The joke is that President Obama does not belong in the convenient pigeonhole of the "black race." First, race is not a definable biological entity. Then, on an individual level, half of his genes come from his black father and half from his white mother. The greatest irony about race is that although science and religion both were used in the nineteenth century to support ideas of European racial superiority, modern studies of the human genome suggest that race, as we conventionally understand it in social terms, doesn't exist as a biological reality. Our eyes are highly discerning of even slight differences between the faces of individual people. We have evolved the ability to instantly recognize acquaintances and even to easily pick out familiar faces in a large crowd. This ability has been under strong selection, because survival long has depended on spotting potentially dangerous strangers and recognizing members of our own family and tribe. There is much more to the genetics of being human than skin color and variation in facial features, but these are easily spotted and used to identify same and not same.

Homo sapiens arose in Africa, our shared birthplace. Stepping away from a tribal outlook, modern humans are all closely related and radiated from African people who lived about fifty thousand years ago. African populations vary more genetically from one another than non-African populations do because our species has existed in Africa longer than elsewhere. All of the populations that left Africa are genetically a subset of a larger African human genetic variation with their own genetic drift added in. Genetic relationships are closer among regionally nearby populations, but the socially recognized "races" do not correspond in any biologically defined way with genetic populations. Other genetic studies show that about 90 percent of variation in humans occurs between individuals within a population on a continent, with about 10 percent of variation accounted for by differences between geographical populations. Racial categories such as "African American" or "Hispanic" are ethnic and historical constructs, not biological definitions, and racial assignments have limited meaning as these groups blend genetically with other populations in a society.

Does anyone remember these facile old classics – the Nordic race, the Jewish race, the Anglo-Saxon race, the Slavic race? It would be better to recognize that there are human populations and to drop the pejorative notion of race with its weighted social baggage altogether. Human populations have both individual variation among their members and differences from other populations. But all this is dynamic. Populations change when new genes are introduced by mating between members of different populations, and even within a population selection for particular traits and even non-selective drift will change the genetic makeup of a population over time. This is evolution. It goes on in us, too, and evolution will not respect simpleminded societal categorizations such as "race."

And on the eighth day God said, “Okay, Murphy, you’re in charge!”

Anonymous

EIGHT

Learning to Love the Bomb

AS I WAS COMPLETING MY PH.D. DISSERTATION IN THE SPRING of 1967, I uneasily awaited my orders to report for active duty in the U.S. Navy. The anticipation was tense because I was a line officer and thus could have been sent to serve in a warship cruising off the Vietnam coast, a depressing thought. The Vietnam War was every day becoming ever more obviously a futile exercise made up of empty victories and doctored body counts. Each night statistics were reported in television briefings featuring confident and heavily be-medaled generals who jabbed pointers at authoritative-looking multicolor charts. No matter how glorious the assertions, effective victories were elusive. The war was a waste of life without any claim to a valid purpose. There were fevered references to hapless countries succumbing to communism like falling dominoes if American troops went home, but in reality combat churned on only to protect the reputation of a president who couldn't admit defeat. The fateful orders finally arrived. The envelope lying on my desk held my future – orders were orders. It seemed like the famous box containing Schrödinger's cat, which existed in an undecided state between life and death. I opened the envelope; the cat was alive. I had been assigned to a research institute at the National Naval Medical Center in Bethesda, Maryland, which had a slot for a line officer. With my assignment to active duty, Beth got a booklet in the mail called *Welcome Aboard to the Navy Wife*. She was incredulous at the instructions about when and how to wear white gloves and, worst of all, that she should respond to the wishes of her commanding officer's wife as her husband did to his commanding officer's orders. Actually, nothing of the sort happened, and we shared social occasions with friends serving similar tours of duty.

The command to which I was assigned was known by the strange abbreviation AFRRI, which stood for Armed Forces Radiobiology Research Institute. This was an unusual organization made up of active-duty military scientists from all three services and civilian scientists working for the Department of Defense. After a few days of settling in, I was assigned to work with a civilian staff scientist, but it was not to either of our tastes. His research appeared to me to be wheel spinning. He wasn't interested in having anyone working with him. Our respective outlooks on doing science made for some friction, but that was quickly resolved. The upshot was that I formally stayed in his lab, but the powers at the institute allowed me to do my own project – anything at all as long as it involved irradiating something biological. In my case it was tissue culture cells. I was pleased I had the freedom to work in this way, but I had not tweaked to my true role in the scheme of things. I was in for a surprise, but that wouldn't come until years later.

At the time I started at AFRRI, molecular biology, cell biology, and developmental biology were together flashing into prominence with an air of making a new biology. I took an evening class in the resurgent science of developmental biology sponsored by a consortium of universities in Washington, D.C., and another in molecular biology offered at the National Institutes of Health (NIH) just across Wisconsin Avenue from the high tower of the Naval Medical Center. These were both exciting in revealing the current state of research in these fields and influenced me to enter molecular developmental biology in my postdoctoral work. I started a project on an early approach using human tissue culture cells to studying how genes regulated cellular action. In keeping with the goals of the place, I irradiated cells regularly and faithfully signed the irradiation logbook. In fact, irradiation provided a useful tool because it affected how cells behaved in their molecular actions. There were no restrictions on my projects, and I was allowed to publish my research in the open scientific literature.

There were other agreeable surprises. The first time I walked out the back door into the brilliant sunshine at lunch time to explore the woods and a pond that lay on the grounds on the Navy Medical campus, I was startled to find myself being loudly hissed at by large, angry raptors in wire cages sitting in the shade of the loading dock. The hawks belonged to a civilian behavioral scientist who was in the process of moving to another job. I never met him, but I became friends with Malcolm Meaburn, another unexpected civilian scientist. Malcolm was a British chemist who worked with lasers. I don't know what he was hired to do or why he moved from England to a U.S. military establishment, but we spent time talking about such things as the chemistry of the early Earth's atmosphere and we published a paper together in *Nature* on reactions that might have taken place. One night Beth and I introduced him and his ten-year-old son to the singing bullfrogs in the pond back in the trees on the Navy Medical grounds. Unexpectedly, Malcolm's son was completely startled and entranced by a something that was simply a childhood commonplace to us, the blinking lights in the trees, the fireflies, which he had never before seen. There are only two species of fireflies in Britain, and they are spotty in their distribution.

If walking around a military installation at night looking for bullfrogs sounds oddly casual, remember these days were long before the terrorism panic hit American life. Neither the Naval Medical Center nor NIH across Wisconsin Avenue was fortified. Both had open campuses. Now, oppressive high fences surround the two campuses and the symbols of government have become all about fear. In 2006 I was invited to NIH to give one of a series of lectures on evolution. The security check needed for an American to enter NIH was more severe than what it took me as an American to enter Soviet-era East Berlin.

I met interesting people at AFRRI and in the larger Naval Medical community. A few were the gratifyingly quirky eccentrics who could somehow flower in a military environment. One of those was a full colonel in the army's Medical Corps who was incensed that people took soup crackers from the little lunchroom without buying soup. He spent hours lurking in a room across the hall trying to catch the culprits – with no luck as he was just too high ranking for invisibility. His actual job was always a mystery. Most staffers, civilian and military, were enormously friendly and helpful. Some had impressive personal histories. One of the nuclear technicians was a survivor of the Bataan Death March and told me quite unemotionally about his three years in a Japanese POW camp.

The modern tragedy of Vietnam, though, was never far off. My neighbor and friend Marvin Goldstein was a psychiatrist at the Bethesda Naval Hospital. We carpooled together. His patients suffered from physical brain damage as well as psychological stress from Vietnam. Marvin's job was treating young patients with tragic brain injuries that resembled the effects of stroke. Others of his patients included Marines trying to avoid returning to Vietnam by working to convince him they were nuts. One morning after he had dropped me off, he drove to the hospital staff parking lot. There he found a patient waiting for his arrival. As Marvin got out of his car, the patient strolled over, hands in his pockets. He looked down at the front of Marvin's car and asked, "Doc, did you know you had a hole in your bumper?" Marvin looked. "No, there's no hole there." The patient pulled out a handgun and calmly shot a hole in the bumper. "Now there is."

Marvin called the Military Police. The patient stood quietly by Marvin's car waiting for them to arrive. He thought he had found the best way to demonstrate his craziness to his psychiatrist, but Marvin thought patients like him were acting perfectly sanely, if a little histrionically. The patients who worried him deeply were those trying to convince him that they were sane so that they could return to Vietnam. A few had developed a taste for war and killing; one had a little sideline business using guitar strings to make garrotes to sell to marines returning to Vietnam. This wasn't the only stress of the time. I gave a seminar at Catholic University in the spring of 1968. As I left the university for home late in the afternoon of my talk, I noticed smoke rising from the Washington skyline. This was the onset of rioting that followed the assassination of Martin Luther King. I drove out of the city without incident, but it was sobering for the next few days to see the capital city locked down under military guard. Armed security has now become a national obsession, but it was not always so.

The AFRRI had its moral ambiguities. It had a research nuclear reactor, a fascinating entity in its own right. Who would have imagined that there was a nuclear reactor operating just off the main street in Bethesda, about eight miles from the White House? Then there was the beast itself. The reactor fuel rods were submerged in a "swimming pool" with water as a moderator. You could safely stand on a catwalk a few yards from the killing radiation and look down at the source shielded by water. It was an unworldly experience to walk around the pool and gaze into the heart of a live nuclear device in clear water that glowed an entrancing blue from the Cherenkov radiation. This phenomenon comes from charged particles moving faster than the speed light is transmitted in water. Electrons in the atoms of the water are displaced and photons are released as the electrons return to their prior state. The blue light was the visible result. Nuclear war was the institute's reason for existence. Bomb tests were celebrated in the seminar room, which had large-format color photographs of nuclear explosions on the walls all around the room. I remember Beth's first visit to my place of work. She quite reasonably gasped at the mushroom cloud photo gallery. I guess we all did on first sight, but we became acclimated to sitting under the atomic art for meetings and seminars.

Eventually it began to seem strange that I was allowed to work in such a loosely constrained way. I would begin to find that much of the other research in the institute was specifically targeted to what seemed like a grotesque and morbid objective. Those were the days of the Cold War, when the possibility of nuclear war with the Soviet Union was always on offer. Thus at AFRRI one of the stated goals was to learn how exposure to sub-lethal and lethal levels of irradiation from nuclear-tipped antiaircraft missiles might affect the will and ability of a bomber pilot carrying out his mission. I would learn that there was a project to study that question using rhesus monkeys. I now and then saw the monkeys, and someone explained to me that first rhesus monkeys were trained to perform certain measurable tasks, then investigators irradiated them with various doses of neutrons. The pulses of neutrons for the studies came from the heart of the nuclear reactor.

The protocol of the primate studies was to study the performance decrement in trained animals after they had been irradiated. At the high end, large pulses of neutrons killed instantly by short-circuiting the central nervous system, a shocking event that I saw once, near the end of my tour of duty. While passing through the reactor workroom I saw a television monitor, which showed a monkey sitting harnessed into a seat. He looked around, seemingly interested but not frightened. The technician told me it was a neutron beam test. "Watch," he said. The monkey suddenly slumped, head lolling. Lower doses of radiation sickened the monkeys and lowered their enthusiasm for performing their assigned tasks. They were encouraged to keep working by being reminded with "mild" electric shocks. Performance decrement as a function of radiation dose was evaluated. From what I heard of the behavioral project, it seemed like misguided, cruel, and scientifically indefensible work to me, but it was cast as leading to at least some protection for flight crews. What I hadn't guessed is that the public objective of the monkey studies was as much a cover for the real project as was the whole other façade of basic research, of which I was a part. As repugnant as the studies with intelligent monkeys were, if the objective of preserving human lives had been true, maybe one could argue that they were defensible. But the unannounced goal of irradiating monkeys was to help design a weapon that would take human lives but preserve property.

The light only dawned a decade later when I arrived at my lab at Indiana University one morning. My graduate students gleefully showed me an article in the January 20, 1978, issue of *Science*. "Isn't this the place you once worked?" one asked. It was indeed. An article written by science reporter Nicholas Wade revealed that the government of India had abrogated the rhesus monkey treaty under which monkeys exported to the United States could be used for medical research but not for military or space research. The reason cited was the monkey business at AFRRI. Someone from an animal rights group somehow found out what lay behind the appalling work being done with the monkeys and had finally publicly revealed the whole story. These primate studies turned out to be a serious violation of the rhesus monkey treaty, with its restrictions of kinds of experiments allowed with monkeys exported from India. *Time* magazine reported the international fuss that blew up, a fuss that had the serious effect of temporarily stopping the import of rhesus monkeys for research on vaccines. A Berkeley-based animal rights group called the International Primate Protection League had reported that the AFRRI studies were the dosage studies for the controversial neutron bomb. This was a real estate–friendly weapon intended to leave the buildings in its target zone intact while it killed all their occupants. The neutron bomb was an ultra-secret program that I knew nothing about while I was on duty. Much was concealed in the open. The rest of us seem to have been embedded to do research at AFRRI as unknowing cover-up scientists. That is something that has remained a disquieting and distasteful memory for me.

First-class stories seem to never die, and this one was no exception. The familiar plot reappeared to my surprise when I saw the 1987 movie *Project X.* There was the thinly disguised tale of the AFRRI monkeys. But it was jazzed up to involve a secret military base in the Florida swamps where chimps were trained to fly simulators and then exposed to radiation to test the responses of human pilots exposed to lethal doses of radiation. Naturally, the hero saves the chimps and gets the girl. The rhesus monkeys had no such happy ending. Nor apparently has there been much change in attitude. As I was writing this part of the book, I learned that NASA, forty years later, was planning to do the same experiments on monkeys again, this time to study the effects of long-term low-level radiation on astronauts in interplanetary space. The plan was to train squirrel monkeys to perform tasks and then, after exposure to low-level irradiation, watch the effects of irradiation on performance over an interval of time. At least this recycling of the study and its gilded space age rationale didn't escape attention and outrage.

There was a strange sort of closure to this story with the death of physicist Samuel T. Cohen in December 2010. Cohen was the inventor of the neutron bomb. In a September 2010 telephone interview for his obituary, soon to be published in the *New York Times* (I thought, what a strange call to get). Cohen opined, "It's the most sane and moral weapon ever devised. . . . It's the only nuclear weapon in history that makes sense in waging war. When the war is over, the world is still intact."

During the first six months of my time in the navy I stayed with Beth's parents in Bethesda and drove to Durham each Friday afternoon to spend the weekend. Living with the dread in-laws? Hardly. I liked Beth's family and enjoyed staying with them for the few months. Courtesy of Wilma I had a lovely complete breakfast every morning. At first I was in awe of Beth's father, Ed, because he held a high position in the congressional staff and seemed reserved. As his office worked on the drafting of legislation with congressmen of both political parties, he had to be circumspect. I was to learn in time that he had a lively mischievous sense of humor and cared passionately about politics. His best flash of humor came in the wake of his visit to AFRRI. He wanted to see the reactor. That was not classified, so I could bring him to the institute as a visitor. When he left, I was immediately summoned into the presence of a panicked second in command. Seems that Ed had signed in by his title, "Legislative Counsel, U. S. House of Representatives." So unknown to me, a "congressman aboard" alert had gone out and pretty thoroughly stirred up the place. I relieved the commander's anxiety with the news that my visitor was not a congressman. When I told Ed about the reaction, he smiled beatifically and said, "I didn't think it would do you any harm."

Once Beth finished the last of her experiments at Duke, she moved to Washington to write her dissertation. We rented an apartment in Rockville, Maryland. When, as not so rarely happened, Beth got bored with thesis writing, she would cook marvelous and intricate dinners. The balconies of the apartment units faced either the parking area in front or overlooked the green and partially wooded cemetery behind. For five dollars extra per month rent, we chose a balcony view of the cemetery greensward. It had in its trees a spectacular diversity of birds easily lured from the relatively slimmer pickings of the cemetery to the rich fullness of a birdseed feeder. We also had a neighbor who took her young daughter out into the cemetery for walks. Seeing them, mother and daughter, hand in hand moving among the grave markers, companionably picking bouquets of the plastic flowers left for the dead, was both charming and creepy, and a constant wonder. The armfuls of plastic flowers that came home from the mother and daughter walks accumulated in great bundles in pots set out on their balcony, not dead, never alive, immune to the seasons.

Nature: a place where birds fly around uncooked.

Oscar Wilde

NINE

On the Road to Chiapas

AS 1968 BLOOMED, WE WATCHED IN DISMAY THE GROWING ferocity of the Vietnam War with the shattering surprise of the Tet Offensive at the beginning of the year, the decline of Lyndon Johnson's hold on the presidency, and the depressing inevitability of Nixon's election that autumn. However, my time in the U.S. Navy would come to an end early in the summer of 1969, and exciting science beckoned. One enticing new discipline bubbled up about that time from the sudden re-awakening of developmental biology: the study of how organisms develop from the fertilized egg to the adult, with the marvelous unfolding of the elaborate form of an organism from the apparently structureless spherical and homogeneous egg. I looked for postdoctoral opportunities in this new molecular developmental biology being pioneered by just a few labs. Developmental biology had made famous discoveries earlier in the twentieth century through ingenious microsurgical experiments but had languished by the 1960s because of a lack of tools to penetrate further into how embryos work. This was about to change in a completely unexpected way as genes became better understood and ways to study them were found.

I applied to labs doing interesting work in development and soon was invited for several interviews. One was at Princeton with John Bonner, who investigated the wonders of how slime mold cells swarm to form a spore stalk – pioneering work on establishing how cells communicate. His lab was on the floor directly above what was then a marvelous exhibit of dinosaurs at Princeton's Natural History Museum (the dinosaurs have since moved on with the transfer of the Princeton museum's vertebrate fossils to Yale). Another research group at a different university worked on the polytene chromosomes of carrion flies raised in barrels of decaying meat, an unnecessary bit of personal heroism it seemed to me. I signed up to work at the Massachusetts Institute of Technology with Paul R. Gross, whose work on gene control in the development of sea urchin embryos struck me as exciting and oriented toward animal development. The arrangements that would lead to my work with Paul were made in 1968, and I wrote a successful grant proposal to the American Cancer Society to fund my salary to support my two years at MIT.

My service in the navy would end the first of June in 1969, and our time in Boston was to start in September. George and Alice Beatty invited us to join them in the next of their almost yearly dragonfly-collecting trips to Mexico. I've never known what the scientific mission was beyond surveying the dragonflies of Mexico. I doubt that they ever formulated a specific research goal, but the trips did provide a structured rationale for going into the natural areas of Mexico they loved best. The collection of specimens was a kind of naturalist's ceremony and a way to document through taxonomy what organisms lived where.

We helped George and Alice fit out the 1949 school bus, which George had patiently turned into a forest green sixties-paleo version of a camper. The thing was completely home built, seemingly around a cranky propane gas–powered refrigerator that had to be absolutely level to cool. This appliance would make parking on uneven ground something of an expletive-accompanied art form once we were camping in the bush. Getting started was a slow affair. George was a great perfectionist and thus a mighty and frustrating procrastinator. As June wore on, we finally had to force him to move by telling him we would not be able to go if it took any longer because we had to start our postdoctoral positions in Boston at the end of summer. Within a couple of days everything suddenly got shifted into the bus. Here a philosophical difference on reading books on an expedition emerged. All of us were addicted to reading, and Beth and I had brought so many books that they had to be tucked into corners of food storage bins and under mattresses. George brought not a single book. He felt that the purity of the traveling experience would be diminished by reading rather than a total immersion in the journey – even the boring parts. "Reading distracts you from absorbing the trip," said George. But I found the time on the road when I wasn't sweating behind the wheel, or looking nervously down into deep ravines as we careened down narrow winding roads, glorious for reading. As the school bus trundled down an endless Texas highway, I opened *The Major Features of Evolution,* by the legendary paleontologist George Gaylord Simpson. There he laid out the theoretical foundations, as he saw them, for connecting the evolutionary record seen in the fossil record with evolutionary processes operating in the living world. I read *Major Features* out of curiosity, but his theme would become one of the major threads of my later work in evolution.

We bought provisions of food and beer, did our laundry, checked for forgotten items, and began the drive. A trip of thousands of miles on a bus capable of a top speed of forty-five miles an hour can be trying, and there were a few unadvertised quirks to our venerable machine itself. This ancient vehicle had a transmission that required "double clutching" when downshifting. That meant engaging the clutch, putting the gears into neutral, releasing the clutch, reengaging, and shifting down. Easy enough to master on a level road, but just a little tense on those occasions when the bus started to accelerate down a precipitous mountain track with the all-too-familiar vertiginous cliff on one side. There wasn't time for fumbling with gear shifting right then. The other interesting mechanical feature of the bus was that the driver's part of the cab was an open portal into the mouth of a roaring, grinding furnace – the massive hot and un-insulated engine. There was no air conditioning. In the humid summer of Arkansas, Texas, and the Mexican lowlands, the waves of engine heat were exhausting. Driving at night was a relief.

Eventually, we arrived at the Mexican border. The customs officers took one look at our unconventional outfit and sidelined us. We sat on a bench in a nearly empty corridor for three hours or so. Now and then someone would appear, ask the incidental question, then vanish again. Finally we found out what was holding things up: not the monster camping bus packed with stuff from floor to ceiling but our seemingly innocuous Grumman canoe strapped open side down to the roof. We had taken it along thinking it would be fun to put in the water somewhere in Mexico. Actually, as its hull was a voluminous dry space, the canoe soon got co-opted for storage, and it stayed that way all through the journey.

Finally, as we sat waiting, yet another customs agent appeared. He asked us just what the boat was. We explained. He returned a while later with our permit to enter Mexico. The canoe jam was broken. It was described on the document as "canoe, aboriginal, aluminum."

Once south of the border, the scenery unrolled from Texas scrub into the unfamiliar and wonderful. We crossed the Tropic of Cancer in high desert and entered the tropics at El Salto, a turquoise waterfall that decanted into a series of downstream basins of pale green travertine, breathtaking thirty-foot saucers where we swam as in a dream. There were no people, just pairs of scarlet macaws that flew noisily overhead and the sound of the water. We continued on to Tamazunchale (which gringos pronounced "Thomas and Charlie") and then across the Sierra Madre Orientale, the Mexican high desert with its painful and treacherous cholla cactuses, where weirdly branched and tufted Joshua trees and the green, tree-tall and slender columns of El Cardon organ pipe cactuses filled the skyline.

In the evening of July 20, we stopped to buy some beer at a scruffy roadside tavern frequented by the drivers of the ornately painted fast trucks whose sudden appearance in the opposite lane made driving the cliff-edged mountain roads of Mexico a sweating palms adventure. The curious tableaus of rusting truck carcasses lying two hundred feet below in one canyon or another made for thought-provoking exhibits should you care to dwell on them as you drove the same road. George and I went into the tavern alone because he felt women would not be welcomed there. I wondered just how welcome we would be, but we found the bar patrons quite cheerful. They motioned us to the television, where we were stunned to see American astronauts on the Moon. *Apollo 11* had landed, and Neil Armstrong had just taken the first human step on another planet. We knew that the moon landing was coming up, but had been on the road long enough to have forgotten about it. As Neil Armstrong's steps left their immortal impression in the lunar dust, we carried our bottles of Dos Equis back across the Mexican dust to the bus. Such are the quirks of memory – the mundane remembered by association with the dramatic.

After that it was endless miles of driving across another novel country of agave farms where fermented pulche, a cloudy alcoholic drink, was made. There were highland markets, volcanic spatter cones, and always, in the middle distance, the perfect conical peaks of the great volcanoes Orizaba (Aztec name, Citlaltépetl), Iztaccíhuatl, and Popocatépetl. All three mountains are over seventeen thousand feet tall, and Popo is frequently active and dangerous to the several million people who live within its historic reach. Mexico and Central America are tectonically active, sometimes violently so. The great Mexican volcanoes rest in a line that lies athwart the country. These beautiful volcanic peaks are the stupendous release vents for the heat and gasses produced by the subduction and melting of the Cocos Plate, which is being destroyed under the crust of Mexico. Clear crater lakes lie, with a deceptive peacefulness, down in the plain, and people fish there, in a half-mile-wide hole blasted out a geological instant ago by an unimaginable volcanic explosion.

Then it was literally downhill from the high desert to the rain forest town of Tuxtapec. We stayed with a longtime friend of George and Alice's, a dedicated physician named Martin Hidalgo, whose family lived in Mexico City. He spent his weekdays in Tuxtapec caring for the villagers. This was tough duty in a lowland climate of unrelieved, day-and-night hot and muggy air, in a town that appeared to exist in a permanent undefined other century. Hidalgo's house had walls of an open brickwork pattern, which allowed for a languid flow of air. It also allowed geckos and other curious beings the freedom of the house. The shower was home to an uncomfortably large but somnolent hunting spider that spent its time resting close to the water taps. No one bothered him lest he be replaced by something worse. A large tamarind tree shaded the patio. There was a popular tamarind-flavored soft drink for sale in Mexico, but it never appeared in the United States. Tuxtapec had a timeless local market where people spread out on rugs medicinal herbs, food, baked clay pots, and clothing. It seemed like an exotic experience to roam through the town market, but it was in fact an everyday event to the inhabitants. Someone had by his stall a large but friendly pet peccary that liked to have its stomach rubbed. As there was no municipal garbage collection, refuse was tossed into neatly walled back yards that held a pig or two. There was no excess of plastic bags in those days, so no non-biodegradable mess to blow in the wind and accumulate in gutters and gardens. The zopalotes (well, actually vultures) adorned every rooftop along every residential street, patiently waiting their turns at fresh inputs of backyard refuse. Seemed to be an effective natural sanitation system, and the rows of beady-eyed zopalotes nicely dressed up the eves of the houses.

Neither Beth nor I had ever been to the tropics, and they were to prove as startling to us as they have to all first-time visiting naturalists from the North. Tropical forests have an immense impact on anyone with eyes open to biology. They are simply lurid, and fecund beyond belief. Ground level has a particular suffused green light from which the multitiered closed canopy foliage above has sucked the brightness of the Sun. Tree boles stand like smoothly tapered Greek pillars; their tops meld with the other crowns of foliage overhead. It is somber, humid, oppressive, close, dank, hot, slippery, spiny, and exuberantly jammed with life in a way that has to be experienced to be comprehended. Every surface is alive. Tree bark, roots, buttresses, and fallen trees are covered with yet more life, vines, creepers, epiphytes, mosses, and fungi that produce a feeling that some describe as a "green hell." For us (not actually having to live in the forest) it was a constant wonder – a continuous "look at this; look over here" experience.

Things are never quite what they seem in such a place. Scrambling out of one streambed, I put out my hand to grasp a green vine but stopped. In the midst of my reaching, the vine transformed itself into the shape of a well-camouflaged tree snake, slender and thin and motionless, hidden in plain sight. Poisonous probably. Many tropical tree snakes are rear-fanged snakes with venoms of varying degrees of toxicity. Their short fangs are in the rear of their mouths, an inefficient system for poison delivery unless they are allowed to chew on a finger for a while. They are un-aggressive, preferring disguise and concealment to attack. Snakes are present, but one sees few. Nonetheless, there are other, more sinister defense strategies by organisms one would never suspect of such shenanigans. The innocuous looking *Cecropia* trees are pioneers in secondary growth. They have an attractive and striking foliage like an umbrella made of a dozen or so leaves radiating from the end of their long branches. Tempting to touch, but not friendly plants. The trees have chambers inside their trunks in which *Azteca* ants live. The ants are fed by glands on the tree, and in turn they swarm out to bite anyone who gets too intimate. I've been bitten by these furious ants just for the offense of carelessly brushing against their *Cecropia* apartment house. The mountain rain forests were lush with tree ferns and rushing clear streams, where large green tree frogs rested on mossy rocks in the spray. Around more leisurely streams, spectacular *Cora marina* damselflies cruised the edges. Their bodies were enamel blue gems in the shafts of sunlight shining through the leaves overhead.

The life of tropical trees extends to the operation of another kind of above-ground vegetable hotel: bromeliads, which are tree-dwelling relatives of pineapples. They have become familiar in the United States in a tame, potted form. Few buyers looking at these clean, manicured greenhouse plants realize that in nature the plants cling to trees in massive unruly clumps. The deep cups at the leaf bases are filled with water, and these thousands of hidden pools of water held by the plants house a wild diversity of small creatures. Once in Mexico we got up into a tree small enough to climb and used a machete to cut off a relatively small bromeliad for dissection. This was a clumsy operation as we four inexperienced botanical collectors struggled to keep the bromeliad upright while avoiding losing fingers to the machete and/or falling out of the tree. Its leaf-base water pools contained small salamanders, mosquito wigglers, pastel blue worms, and damselfly larvae with their three feathery tail gills.

Some of the damselflies whose larvae live in bromeliads are the largest and weirdest in the world. They include *Megaloprepus,* the helicopter damselfly, with a five-inch-long abdomen. Its wings are dark, but each wing has a white spot near its tip. These damselflies don't cruise the sunlit edges of ponds looking out for gnats to eat and females to mate with like our familiar small brightly colored damselflies. The *Megaloprepus* are the most bizarre of jungle hunters. We looked for them once in the right sorts of woods, somewhat unpleasant and lightless ones with lots of spiderwebs in the understory. There in the gloom were what appeared to be groups of four whirling, white insects hovering innocuously. But these were a deception. The white "insects" resolved into white dots at the wing tips of the hovering helicopter damselflies – a disguise apparently to confuse birds. The nearly black damselflies with dark wings are themselves nearly invisible in the gloom. Why hover in that apparently aimless way? It would seem that it allows them to better focus on and snatch up unsuspecting spiders from their webs, which we saw them do with finesse. In feeding, they are as fastidious as one can be, eating a raw spider and carefully plucking the spider's legs off before settling in for the feast.

I wonder how much of the forest of diverse life remains. Clear-cutting of the hills was already in progress in 1969 by small farmers who could only get a season or two of crops from a cleared plot before erosion stripped the soil away and deposited it as mud in the formerly clear streams. The human population of Mexico has more than doubled since 1969, from about 47 million to 109 million people in 2008. The birth rate in the 1970s was 3.4 percent. The impact of that inexorable population growth on nature and the tropical forests of Chiapas seems to have been relentless.

Somehow George and Alice had managed in their trips to Mexico to befriend an astonishing variety of people who had nothing to do with entomology. We never discovered just how they had done that. One was Martin Hidalgo, who put us up in Tuxtapec and whose family would so generously house us in Mexico City. Another was the internationally recognized Mexican conservationist and author of several books on birds and reptiles of Mexico, Miguel Alverez del Toro, who operated a modest but famous zoo in a rain-forest setting in Tuxtla Gutierrez. He showed us around the zoo, which he had designed to create a natural environment for the animals but which confined the human viewers to caged walkways. The zoo focused on the animals of Chiapas and contained some remarkable animals I had never seen before. The deep woods tapir was shy but willing to exchange looks with us before ambling off. He had the distinction of being an extraordinary and lovely living fossil of the Eocene, about 40-plus million years ago. Tapirs are primitive three-toed relatives of the horse, and they proudly bear odd short trunks on their noses. Another notable ancient was the venomous Mexican beaded lizard, a close relative of the notorious Arizona Gila monster, but large adult beaded lizards are about twice the size of Gila monsters. Unfortunately, beaded lizards are rare and endangered. These were the only ones I've ever seen. Again, this is an old lineage in evolutionary history. Fossils from the 47-million-year-old Messel fossil pit in Germany include a lizard similar to the living beaded lizard, showing that these unique reptiles now restricted to the southwestern United States and Mexico were once widespread. Their venoms are important subjects of molecular evolution studies that show how salivary proteins evolved into venoms. Finally, the zoo had a couple of mystical, seldom seen jaguars – leopard-sized cats sleeping peaceably in the afternoon heat.

These were reasonably biological contacts for George and Alice to have made, but how did they get to know the operator of a Coca-Cola bottling plant, who filled our bus water tank with pure water? We were grateful for that stop, which gave us a few days of not having to drink water that tasted of Clorox bleach, our water purifier of choice south of the boarder. Where did they meet the engineer who managed the power plant associated with a dam? As hardy naturalist campers, we could not shower every night. When we could, it was delicious. At one of those moments of hard travel in the heat, when a shower would have been most desirable, we just happened to turn into the road to the power plant where George's friend, the manager, worked. He welcomed us with open arms and allowed us all to have a shower in their facility. Then, finally, there was the cacao plantation. Once again George knew someone – a cacao grower who lived conveniently nearby. We stopped to visit and were welcomed to make ourselves at home looking for insects under the trees. These are understory plants, and it is a gloomy dark beneath their low branches, with only a few feet of headroom. So we crept around under his cacao trees more or less on hands and knees. There wasn't much insect life, but it was cool under the cacao, and the trees themselves were strange. The small pink and white flowers emerge on stems directly out of the main tree trunk, a rare form of flowering called caulifery. Cacao flowers are pollinated by flies or by the farmer and develop into seedpods about two-thirds the size of a football. At harvest the pods are cut open and the seeds and pulp are removed to a fermentation vat. This step is important to the flavor of the chocolate. If not fermented, the taste of the chocolate will be bitter. Fermented seeds were dried in the sun on canvas sheets. The labor of the process and the ratio of bean to yield of chocolate tell you why it's such expensive stuff. It takes a few hundred beans to produce a pound of chocolate. I must by now, on my own, have accounted for a couple of million cacao seeds.

Occasionally our Mexican lowland forest adventures slid sharply from the inspirational into the utterly silly. In the middle of one night in camp, Alice awoke and said, "George there's something crawling on my back." I heard George sleepily pronounce, "Go back to sleep Alice. You're just imagining it." She was insistent. Needless to say, we all were now awake. I shone a flashlight on her back. There, just above the edge of her nightgown, was a pale little scorpion about an inch long. George piped up all too honestly, "It's only a little scorpion." Alice screamed. Bedding flew. Flashlight beams filled the bus. The scorpion vanished. Now it was loose somewhere in our living quarters. Like demented midnight housemaids, we stood there in the rain-forest dark by the door of the bus shaking out all the bedding. We searched furiously everywhere by flashlight for the scorpion, in sleeping areas, shoes, and storage bins. It was completely fruitless. We finally tired of the hunt and nervously turned in. Alice was sheepish about having betrayed the unwritten code of the entomologist to fearlessly love and respect all bugs. None of us slept easily for the next few nights – each of us waiting for that creepy tickle of tiny feet inside a shirt or on bare skin.

After a few days of life in Tuxtapec, Martin Hidalgo drove Beth and me to Mexico City for a short visit. We stayed with his family, and they took us to the legendary ruin of the great pre-Columbian city of Teotihuacán. This magnificent city was found abandoned by the Aztecs when they came to the country. I had never seen an ancient Mesoamerican city, and this one was outsized in a Hollywood sort of way. I'm sure that the builders intended a setting for a thoroughly vulgar display of power combined with a splash of public human sacrifice. We entered the grand avenue by way of a temple extravagantly mounted with Quetzalcoatl serpent heads, evidently an earlier temple covered with newer construction by the Aztecs. Ahead were the massive gray Sun and Moon pyramids. In their prime the pyramids would have been garishly painted in bright colors. At the end of the avenue was the Teotihuacán museum near the Pyramid of the Sun, where some of the centuries-old Quetzal mural paintings were displayed, a hint of their brilliant ancient pigment remaining. Teotihuacán was at the height of its power from 150 to 450 AD and exerted profound influence over much of Mesoamerica, even to the Maya. It covered over ten square miles, and its population may have approached a quarter of a million people. The other storehouse of the Mexican past was the Museum of Anthropology in Chapultepec Park, one of the great museums of the world, with holdings such as the twelve-foot-diameter Aztec Sun Stone excavated in the eighteenth century in central Mexico City, an enormous "football player" Olmec head from La Venta in Veracruz, and the reconstructed tomb of a seventh-century AD ruler of Palenque with his intricate jade funeral mask. This ruler would be later identified as Pacal I.

We rejoined George and Alice in the Tuxtlas, a mountainous area in southern Veracruz state. As we moved farther south in Mexico, other life forms appeared. We were surprised by our first "Jesus Christo" lizard, which rocketed off across the water when we disturbed it in the bed of a shady stream near Lake Camacho. Herpetologists know this lizard as the basilisk. These harmless basilisks in repose look like leggy and slender but otherwise ordinary lizards up to about two feet in length. However, they have a few dramatic anatomical features – a notable crest on their heads and long gangly hind legs. When pursued, they run upright on those long hind legs in frantic imitation of small dinosaurs. They can even do this over the surface of water – hence their Spanish name. We had them race our bus down a dirt track making their dinosaur run, and we had them startle us when we walked down forest streams when they would shoot off their basking logs and make a sudden noisy bipedal dash across the surface of clear pools in the streambed. The physics of how the Jesus lizards manage to run upright on water was solved by researchers Tonia Hsieh and George Lauder, who in 2004 made movies of them running across a glass-walled tank of water in a lab. The lizard's feet don't stay on the surface as if the water were a skating rink. Instead, their legs act as pistons that displace the water such that their legs are below water level in a shaft of air and their feet are pressing against the water at the bottom. They have to move quickly before the shafts of air close or they will get a dunking. Their feet also provide balancing lateral thrust so they can keep upright through their run. After ten or fifteen feet of aqueous sprinting, the lizards run out of steam. They slow, and if they haven't reached the other side of the stream in their run, they have to ignominiously swim the rest of the way.

For most our time in the Tuxtlas, we parked on a little island in the Rio de las Ranas, an idyllic jungle scene, peaceful with people wading the nearby ford in the stream now and then. Still, the possibility of meeting up with a fer-de-lance viper in the tall grass made us highly cautious about where we put our feet. Walking in the water was best. The streambed was a shallow, cool, and shaded tunnel through the forest with hummingbirds visiting the occasional clumps of bird of paradise flowers overhead. They sounded like helicopters in the dense foliage but were hard to spot. There Beth and I studied an unusual species of damselfly, the *Paraphelbia zoe,* a species in which the males come either with a black band on the wings or with clear wings. They were mating with females with clear wings. We wondered if these males represented one species or two. We marked the wings of mating pairs and showed that the same female could be re-caught mating with a second male of the other wing color pattern. We never published the data, but a paper was published in the journal *Evolutionary Ecology* in 2008 by Romo-Beltrán and colleagues discussing this phenomenon. What we privately discovered is now well known. Sadly, this species like many others is threatened by habitat loss.

We stopped at Villahermosa in Tabasco to see the ten-foot-tall Olmec heads carved from basalt. We were allowed to camp in the nearby municipal park for a couple of nights – not a campground, just us at the edge of a shady park. This location would supply my day of inadvertent adventure. Some items had been pilfered from the bus, so I went to the municipal police station to file a theft report. I didn't expect anyone to investigate anything, but in order to collect insurance on theft, one needed a record of a police report. The cops were having a quiet day, so instead of just giving me a copy of a theft report, they assigned three bored detectives to the "case." First they showed me around the jail. That was an eye opener. The building enclosed a three-story open dank space with a surrounding catwalk, on which guards armed with rifles lounged. The prisoners below were visiting with their families, who had brought food – quite casual. We then got into a police jeep and careened through the streets of Villahermosa to look for the thief – a twelve year old befriended by Alice. We conversed in a makeshift sort of way. I spoke with the kind of Spanish proficiency sufficient to buy tomatoes at the market. They spoke no English at all. Still, the detectives and I had a fine time fighting crime together.

Villahermosa houses a unique collection of Olmec heads in a public garden, La Venta Park. These extraordinary objects, giant heads carved of basalt that look like they are wearing football helmets, were exhumed at the archeological excavation of La Venta, a site dating back to nearly 1000 BC. The Olmecs built the first Mesoamerican civilization and influenced the culture of all that followed. Weirdly, the great Olmec heads in Villahermosa would be vandalized in 2009 by a trio of evangelical church pastors, two Mexicans and an American, who, claiming the heads were satanic, poured oil, grape juice, and saltwater on them. I don't know if they meant to damage or to somehow anoint the sculptures with this peculiar mix of ingredients.

Culturally, the most stunning part of Mexico was Chiapas. We drove with the goal of reaching the ancient Maya city of Palenque, crossing mountain ridges high enough to strain the old bus engine. We crept higher and only entered into the mountains once it was night, making for rainy and tense driving. The darkness and mist provided an unreal scene of the elephant ear–sized leaves of *Gunnera* plants appearing and vanishing in our headlights as we slithered on the muddy track. Then we hit the real professional-grade mud near the village of Tapilula. Worse, we had a flat tire and spent the night in the bus stuck in the middle of the road. This camping spot was precarious because we would be invisible to anyone careening along that road. But there was no careening – only fog and muffled silence. When we awoke in the morning, the rising fog revealed a line of vehicles embedded in the mud on the road behind and ahead of us. We sat on a grassy hill and watched as big yellow Japanese road graders pulled all the night's accumulated cars and trucks out one by one. It may be so to this day, because the graders couldn't do much roadwork until the night's catch of trapped vehicles was removed. Only after we had sat happily on a grassy slope to comfortably watch the pulling of the line of vehicles off the road did we get back to the bus – once it too was pulled clear. The graders deposited the bus nicely on the roadside beyond the slough. It was then that we faced the sad truth that we had a flat tire to change. This tire was one of the double wheels at the rear end of the bus, and it was lacerated by large nails bent like iron talons that firmly attached a wooden plank to the distressed rubber. George and I were the designated naturalist mechanics. Our morning's entertainment was to jack up the bus and remove the shredded tire and the heavy gumbo that filled the wheel well. Luckily, the flat was in the outer tire on the axle, but still it was more mud wrestling than mechanics. We each wore a proud coating of Mexican upland mud as we got rolling again.

Chiapas is Maya country. One might imagine that the Maya and their language and culture should have passed from the living world with the abandonment of their cities between 800 and 900 AD, leaving a mystery as to who built those vine-grown pyramids. In fact, a few million speakers of Mayan languages still live in Mexico and Guatemala. No matter this inconvenient little fact, a whole genre of writing invoking space aliens as the mysterious builders of the actually very human Mayan monuments remains ever popular. We were to meet guards at archeological sites who spoke sometimes only Mayan and who, with faces that resembled their forebears of a thousand years before, looked like the spectacular Mayan profiles carved centuries ago on the steles they watched over.

Palenque itself was an evocative and stunning place. The setting in which we first saw it was in the early morning as the mist was rising and the site was still empty of people. We climbed the stone stair way to the top of the Temple of the Inscriptions and looked over the excavated and restored part of the city at temples and palaces standing in lawns incongruously well maintained in the great expanse of rain forest. Tree-covered mounds spoke of more pyramids, mysteries still unexcavated. A pair of white horses grazed in the midst of the intense green of Palenque in front of the main palaces. It was a magical, ephemeral view, and in my memory it looked like it might have been a movie scene created for *The Lord of the Rings,* but horses roaming Mesoamerican cities were an impossible anachronism. The Temple of the Inscriptions was not as previously thought merely a temple; it was also the tomb of the great king Pacal I (625–683 AD). I don't think his name or anything about his history was known when we stood atop his pyramid. The Maya glyphs had not yet been decoded and the Maya were still without a written history. A steep stone stair descended into the void, down to Pacal's burial chamber.

The decipherment of the glyphs the Mayans left in their ruins and on their monuments was an enormously difficult feat. The first to be read were numbers, which led to the early-twentieth-century cultural interpretation that the Maya were improbably a peaceful people, their pyramids seen as observatories for dreamy astronomers and the calculators of calendars. The big breakthrough in understanding the inscriptions of Maya rulers and their actual interests came in the 1970s and resulted in the first reading of a list of Maya kings. The glyphs recorded history; the great ruler Pacal was revealed. Subsequent work resulted in a more realistic, less ethereal and peace-loving view of the Mayans. They were a vibrant and creative people, but not all that different from the rest of humanity. Their rulers were not dreamy astronomers. Like rulers elsewhere and elsewhen, they were vain, warlike, and grasping, and their monuments celebrated those virtues. Their calendars and carved stone stele recorded the dates and deeds of kings, especially military conquests in which the bloody sacrifice of prisoners featured prominently. Palenque, like other classic Mayan cities, died in the interval between 800 and 900 AD.

The Mayan people and their folk culture and languages survive to this day, but clearly there was a break in Mayan history with the loss of the great lowland cities and Mayan writing. Each generation of anthropologists likes to apply the worries of our own times to these mysterious great cities abandoned in the jungle. Environmental degradation because of too-high populations and concentrated agriculture as well as warfare have been paraded as likely possibilities. In 2003, climate scientist Gerald Haug and co-workers suggested one likely root cause for the environmental woes suffered by the Maya. Their data revealed that during the interval that saw the Mayan collapse, there was a repeated failure of regular rains and agriculture. Droughts were the result of fluctuations in the position of the climate system that generates rains in the tropics. Once the demanding sacred priest kings failed at placating the gods, their tenure became shaky. Environmental stress culminating in crippling wars would have done the job of abandonment of the cities effectively by beheading the royal dynasties that built the temples and maintained the written records. Glory finally became too expensive, and everyone faded back into a slower paced village life.

The Chiapas mountain city of San Cristóbal de las Casas has notable and pretty colonial buildings, but the impression of the town I carried away were of the vibrant people at the market – the old man sitting grinning on the stairs by his glossy black turkeys with their bright red wattles, the woman standing by a cul-de-sac market stall with coil after coil of hemp rope balanced on her head, the Chamula man striding determinedly down the street in full white traditional costume, including a flat white hat with vibrantly colored ribbons cascading from its crown and an ornate leather shoulder bag. The men had bare legs and the look of great long-distance walkers. The area's rural people dominated the weekly market. The variety of products they brought in from the country to the market or bought there was enormous: live turkeys, chickens, rope, clothing, sheepskins, vegetables, grain, prepared food, pottery, and machetes. Their large water jugs and machetes were regionally distinct.

Mexico was a simpler and far less dangerous place in 1969 than now. The drug cartels to come were unimaginable at the time. The country was generally friendly and safe. Still, some of the local Maya people who lived in the highland region surrounding San Cristóbal did not like outsiders. A few people who had come into town for the market were obviously hostile to cameras. We were careful not to offend, taking pictures of people only with permission or from a long distance. Normally we would have camped in the country, but here it was dangerous to camp anywhere outside of town. A party of campers in the country had been killed a short time before we arrived. These were not American or French tourists, but Mexicans. We took it as a bad sign when Mexican citizens were killed in their own country as outsiders. This xenophobia and conflict with the Mexican government was to later flame into the violent Zapatista rebellion of the early 1990s.

The town's shops were already geared to tourists but offered some interesting local items such as amber from the Simojovel mine, which lay in a remote part of Chiapas. This amber is Miocene in age and is famous for containing fossil insects. One shop had a couple of polished amber necklace bobs shaped like hearts that contained large insects and one a scorpion. They were five dollars each. We knew that it's easy to fake fossils in amber using modern insect road kill, but at that price there wasn't much to lose. Once we got home and I had a chance to show them to an expert insect paleontologist, sure enough, they were easily detected fakes when one knew what to look for. I still have them and treasure them for what they show about how amber fossil fakes are generated for the tourist trade – and just as souvenirs. Chiapas would be the farthest reach of our trip. We returned through Oaxaca, crossing the Isthmus of Tehuantepec to the Atlantic.

From there we started our return to the prosaic. We covered familiar ground quickly but took time to visit the ruin of El Tajín, in Veracruz near Atlantic Coast, our penultimate event on the road back to Texas. This is a well-maintained site, and it was empty of visitors the day we were there. The lonesome guard occupied himself following us around making sure that we didn't commit any indiscretions, like using a tripod when taking pictures. Construction of the ceremonial buildings at El Tajín began about the first century and peaked in the interval of 600–1000 AD. This corresponds to the time span of the peak and fall of the classic Maya centers to the south. The buildings of El Tajín show the cultural influence of Teotihuacán and were certainly grim enough. The famous Temple of the Niches is heavy, graceless, and gloomy, in keeping with the bloodletting that went on there. There are some spectacular carved panels showing such uplifting events as the sacrifice of a living ballplayer with a razor sharp obsidian blade. This is a funny concept to modern Americans, who think of ball players being paid bonuses or traded to other teams as measures of success or failure. The city thrived to the thirteenth century AD, when it was conquered and largely abandoned. A dwindling population continued to live in El Tajín for a while but had vanished by the time the Spanish conquistadores passed through in the early sixteenth century with their desire to do good works and hunt up a little gold on the side. The outcome for the native people was a familiar one – guns, horses, steel, and the invisible smallpox virus. I thought the place was dark and depressing, but that may have been due to the heavy overcast sky. This theater of former human sacrifices might have looked more enticing in bright sunlight.

The nights we spent along the Gulf Coast in the full steam bath of summer as we drove north were the worst of the trip. We tried to sleep, but mostly we sweated and tossed through nights built of heat, humidity, mosquitoes, and penetrating reek of oil refineries. We made our last stop in Mexico in the border town of Matamoras, which is just across the Rio Grande River from Brownsville, Texas. Our ultimate act as departing naturalists was to shamelessly roam the tourist market and listen to mariachi bands. In Brownsville we stayed in the luxury of an air-conditioned motel with an unoccupied swimming pool. We also satisfied our sudden and inexplicable gringo craving for hamburgers.

The trip back to State College took a meandering path. We first stopped to visit the people who were safeguarding our developed slides. In those pre-electronic photography days, we would mail the rolls of undeveloped film to Kodak from Mexico and had Kodak mail the developed slides to one of George's friends in Texas. He was a frog and toad enthusiast who had bound his favorite field guide in marine toad leather – nice and soft, actually. We then moved on to Tulsa, Oklahoma, to visit Alice's relatives, who were enthusiastic about taking us to gaze upon their brand new university. George and I knew what that was all about, but Beth and Alice didn't. So we found excuses to stay home, and they went to have the culturally jarring but eye-opening experience of the full Oral Roberts University campus tour. After seeing Roberts's singular career develop over the next few decades, I thought that maybe I should have gone along too.

Roberts claimed that the place had been built on God's express order. Note, however, that Oral Roberts University was not named for God but for the great faith healer himself. That was only the beginning of God's whispering commands into Roberts's ear. In 1977 a nine-hundred-foot-tall Jesus (why not a thousand feet?) appeared to him and ordered him to build a medical center. The year 1987 brought an even more spectacular twofer. The more profitable of these events happened when the ever-brazen Roberts was told by God that he would be "taken home" unless he could raise eight million dollars. He sent out the word, and the faithful sent their money to save their man from being taken to heaven. The same year *Time* magazine reported that Roberts claimed to have brought a dead child back to life. In 2007, Robert's son Richard resigned as president of ORU in a scandal that involved the conversion of much earnestly solicited godly cash into personal vacations and luxuries. Perhaps there is no culture stranger than our own.

After all, it is as respectable to be a modified monkey as modified dirt.

Thomas Henry Huxley

TEN

The Masked Messenger

WHEN BETH AND I RETURNED TO STATE COLLEGE AT THE END of the long, anticlimactic journey back from Mexico via Oklahoma, we packed our car full of our possessions and set off for Boston to start postdoctoral work. We moved into the entire middle floor of an enormous frame house in Watertown, built about 1900 by a prosperous dentist who had an expansive family to accommodate. Our only pet at the time was a gopher tortoise we had rescued from the highway in Oklahoma. He was charming in his own reptilian way. We'd offer him lettuce and he'd go for it. Once a bite was taken, he'd drift off absentmindedly, lettuce forgotten until he'd catch a glimpse of the green. Then he'd frantically charge back for another bite. We found him a good home with a herpetologist. I settled into Paul's lab, and Beth started as a postdoc in biochemistry at Tufts Medical School. Boston was an entirely new experience. We quickly found the Boston Museum of Fine Arts and the Isabella Stewart Gardner Museum. Both had wonderful collections and were places we could afford to visit on postdoc salaries in an expensive city. Hal White, our friend from Penn State, was now a postdoc at Harvard. His wife Jean was a graduate student at Brandeis University. With them we discovered close-by canoe and swimming streams in New Hampshire and the wonderful Ponkapoag Pond in the Blue Hills nature preserve.

Our landlady, Mrs. West, lived next door in a large white house of the same vintage as ours, with an almost equally antique bulldog named Jazzy. Our daughter Amanda was born at Boston Lying-In Hospital (now called the Boston Hospital for Women) the next summer. It was in that house that she would plunge both hands deep into her first birthday cake.

Who knew they would do that? Her first word at close to a year old came out of her spontaneously as she sat with me looking out the window at Jazzy the bulldog. It was not mama or dada; it was "dag." For years after, although she lived in the Midwest from the end of her first year on, she would claim well into middle school to have a Boston accent.

Through Hal White, I got involved in the Cambridge Entomological Club, which was founded and guided by Frank Carpenter, a famous paleontologist at Harvard working on the evolutionary history of insects. Frank once took me into his office and showed me the fossil wing of a Permian (about 270 million year old) dragonfly imprinted in exquisite detail on the surface of a slab of pale pink siltstone. Living dragonfly wings are an inch or two long. This fossil wing was about a foot long and preserved the complex pattern of the veins of the living wing. The whole insect when alive was the first great aerial predator on Earth. Carpenter had discovered the wing while crouching to wiggle through the strands of a barbed wire fence late on a blazing, thirsty Kansas afternoon. This was the magic hour for finding fossils, a low angle of sunlight that makes every little variation in a rock surface stand out. Fossils on rock surfaces are often invisible at midday, but become obvious in early morning or late afternoon. The Entomological Club met an evening a month in a conference room in the biology building at Harvard. Sitting around a large table, we listened to one another talk about exotic bugs. Some faculty attended, but most presentations were by graduate students. Harvard graduate students in entomology didn't just work on humble local subjects. Their talks featured exotic projects such as studying massive colonies of communal spiders in South America or crawling into ice caves in Idaho to collect ice beetles that can't live above refrigerator temperatures. Hal and I made a report based on our more modest travels.

Hal had collected *Williamsonia lintneri,* a rare dragonfly confined to New England swamps, at Ponkapoag Pond. This was the last genus of North American dragonflies for which the nymph had never been described. We thought we should look for the larva in the spring, when it would be leaving the water to undergo metamorphosis. We caught some nymphs getting ready to emerge and took them home to watch in comfort. Sure enough, the same day one of them climbed out of the water up onto the stick we had standing in its jar as a perch. Beth and I watched the trim, dark adult dragonfly come out from the larval skin and inflate itself on our coffee table as we ate lunch. Having the adult made the link proving that the larva was indeed that of *Williamsonia* and not some embarrassing mistake in identification. Hal and I presented a joint talk at the Entomological Club and jointly wrote the description for the journal *Psyche* in 1970.

While working on the paper, I processed the photograph of the nymph that we would use in the manuscript in Paul Gross's darkroom at MIT. Paul saw me emerge with the prints. He was curious. I told him what they were for. Paul was the master of serious looks. He looked balefully at the photographs and gave his pronouncement: "Don't let anyone around here know that you do this kind of thing." The Department of Biology at MIT was to its toenails hardcore molecular biology. Dragonfly larvae would have been too nineteenth century for anyone to bear. A few years later a friend interviewed at MIT for a faculty job. She asked a group of graduate students if they had taken any classes that didn't have the word "molecular" in their course names. The students thought a moment, and one said brightly, "Oh yes, virology" – the study of virus molecules.

It was through the club that Beth and I would eventually meet the legendary paleontologist, chronicler of the evolution of the fantastic insects of the coal age, and student of the origin of insect flight Jarmila Kukalova Peck. She had been a postdoctoral fellow with Frank Carpenter but had returned to Czechoslovakia just in time for the Prague Spring and the subsequent Soviet invasion and repression in 1968. I arrived in Cambridge after she had gone home. Things did not go well for her there. Stuart Peck was a graduate student and a member of the Entomological Club. After a few months, Stuart left for Czechoslovakia. We then heard that Stuart and Jarmila had gotten married, and he had been able to get her out – a dramatic and romantic rescue indeed. They have lived in Canada ever since. It was in this period that I started an informal discussion group in evolution populated by a few postdoctoral fellows from MIT and Harvard. We met monthly in our apartment. The group persisted for only a few months, but it helped spawn the kind of ideas that come from talking with scientists passionate about a variety of interests.

Our time in Boston was one of powerful ferment in American universities, largely as a result of a festering discontent with the Vietnam War. I first ran into it directly when I was still on active duty and went to MIT for my interview with Paul Gross. I traveled in uniform because that was much cheaper as the airlines offered a generous discount to military travelers. I spent the night with Hal White, who I met up with on the Brandeis campus where he was playing softball. As I paid the cab driver, he remarked, "Captain, I wouldn't go up there dressed like that if I were you." Brandeis was a hotbed of antiwar activism, but I didn't think campus activists maintained a rapid reaction force, and I was right. The worst I got was someone in the middle distance shouting "Killer" at me. It may have helped that Hal, who is well over six feet tall, was carrying a baseball bat and thus looked like an impromptu bodyguard – making for an absurd stroll across campus. Through 1970 and the invasion of Cambodia, feelings were heated at MIT, even to the point of some vandalism against the labs of faculty seen as reactionary. As a result, during the worst of it, I carried my research notebook home at night.

There were regular antiwar demonstrations at Harvard Square. The cops were marshaled on the streets to oppose the demonstrations, and violence broke out frequently. While I was going home one afternoon, we were detoured around the square. On the side street we were directed down, I saw a group of police in riot gear armed with axe handles lined up in front of their buses getting ready for their afternoon's work. There was a curious rumor passed along by some of the participants, with whom I talked, that people whom they didn't know, who were not members of their organizations had triggered the violence at demonstrations. This kind of talk of "outside agitators" was dismissed as leftie paranoia at the time, but later, with the disgrace of Nixon, it emerged that his ever-creative henchmen had sent out hired hands to produce violence at antiwar events, which would make bad press for the antiwar movement.

Once I started in Paul Gross's lab, life became the hunt for the elusive masked messenger. Paul took the messenger RNA hypothesis that French biologists François Jacob and Jacques Monod had developed for gene expression in bacteria and extended it to developing animal embryos. Jacob and Monod's brilliant – and correct – idea was that genes encoded as DNA governed the activities of cells by being copied into messenger RNAs, which associated with sub-cellular "machines" called ribosomes to direct translation, the process of protein synthesis. Their revolutionary theory won Jacob and Monod a Nobel Prize in 1965. In animal cells, which are more complex than bacterial cells, the messenger RNAs must leave the nucleus before translation by the ribosomes into cellular proteins. Paul wanted to see if messenger RNAs were involved in animal development as a way to study how genes regulate the development of an organism. Up to this time no one had investigated embryonic development as a phenomenon of gene action. Paul used sea urchin eggs and embryos. Sea urchins are spiny relatives of starfish. They look like animated pincushions, but with their sharp spines pointing outward. They release glassy eggs into the seawater, where they are fertilized. Under the microscope their larvae are seen as actively swimming and elegant nearly transparent creatures with long arms supported by a finely sculptured skeleton. In Paul's lab I would not yet get to research questions in evo-devo, but among the sea urchins I would find the elegant research organism that I would use for the rest of my career.

Sea urchin embryos were made famous in the nineteenth century with the first observation of an egg in the act of being fertilized. Later they would contribute to our understanding of how asymmetry arises in development and how cells come to differ from one another. The pioneering Swedish embryologist Sven Horstadius did elegant microsurgery in which, by moving individual cells in an embryo, he could show that some cells controlled what tissues other cells would give rise to and how the embryo organized itself. He did not have the tools to explore further. That revolution would depend on the importation of new ideas and techniques from another discipline, molecular biology – as is often the case in science. Sea urchins then became important subjects for the new molecular developmental biology because their embryos could be easily cultured synchronously in the large numbers needed in those days to extract RNA and proteins.

Paul originated the concept of "masked messenger RNA" as a means by which mRNA could be present but not actively translated until fertilization. Unfertilized eggs made few proteins, but as soon as the eggs were fertilized, protein synthesis rose rapidly to high levels. Paul's lab had been able to show that the rise was independent of new mRNA being made. Eggs contained mRNA before they were fertilized and did not have to make their own mRNAs until several hours into development, when their own genes would take over the rest of the process of making the animal. The "masked messenger RNAs" were stored in the egg until it was fertilized. Then they would somehow be released and translated to produce the proteins needed for the beginning of development of the embryo.

I took on the challenge of trying to identify one of those messengers as an approach to finding out what the "masking" might consist of. I chose to study the mRNA for tubulin, the protein that makes up the mitotic spindle that separates the chromosomes during cell division. Newly fertilized sea urchin eggs make large mitotic spindles for their repeated rapid cell divisions and thus had to contain a lot of tubulin. We reasoned that this protein potentially should be synthesized in large amounts from stored mRNAs. A simple way of purifying tubulin had been published. I used the method to show that I could isolate the tubulin protein and show its synthesis by incorporation of radioactive isotope-labeled amino acids introduced into the embryos. I was then able to do the experiments that would clinch the idea that mRNA for tubulin was indeed stored in eggs and translated into protein after fertilization. This was the first time a maternal mRNA had been identified. I was also able to show that the maternal tubulin mRNA was housed in the egg cytoplasm for storage, not the nucleus. My work in Paul's lab would be productive and satisfying because it allowed me to learn molecular biology as well as a great deal about the classic embryology of these animals. This was a stimulating scientific environment, occasionally enlivened by Paul's homilies on scientific integrity.

MIT was an exciting place to be in the late 1960s because it was a powerhouse of the new molecular biology. A number of my contemporaries in Paul's lab later went on to make stellar contributions to the molecular study of development. The most electrifying discovery during my time there was of reverse transcriptase by one of the young faculty members, David Baltimore. I remember him elatedly going down the hall telling people that he had broken the "central dogma." By that he meant the understanding that information always flowed from DNA to RNA to protein. Reverse transcriptase is an enzyme produced by viruses in their life cycle that copy RNA to make DNA – and thus violate what was thought to be the only way genetic information could flow in a cell. It was a stupendous discovery of enormous theoretical and practical importance and won Baltimore a Nobel Prize a few years later.

During that era there was an accelerating drift of prestige and resources from organismal biology to molecular biology. That revolution produced a denigration of organismal, evolutionary, and ecological biology, with substantial effects on what kinds of science get funded well. It also produced some laughable phenomena. The most entertaining piece of theater happened when a visiting postdoc from another institution, who was working on the nervous system of the nematode *C. elegans,* came to MIT to present a departmental seminar. This small nematode worm would become a founding organism in the "model systems" approach to developmental genetics. The idea of model systems is to concentrate research in biology on a very small number of species that are easy to culture in the lab and can be used for genetics. That has worked well for the study of development, but it has reduced the stable of organisms studied and thus can run at cross-purposes to much of evolution and ecology. *C. elegans* has only eight hundred cells total in its body and only a few nerves in its nervous system.

On the day in question the guest speaker was talking about mapping all these nerves and their connections – an important step in more detailed studies in how the nerves work. The speaker had done his mapping by serial sectioning of a complete worm. Thus the whole worm, including its nerves, was cut with a diamond knife into "serial sections" of many fine slices like a microscopic sausage. The images of the tissues in each slice could be reassembled to give a three-dimensional map of the nervous system. It was lovely research. The molecular biologists convened a faculty meeting right after his talk to discuss the possibility of hiring him. I remember Paul returning from the meeting shaking his head. The molecular biologists wanted to hire the day's speaker because they thought he had invented serial sectioning, a technique that had in fact been devised in the nineteenth century, which Paul, I'm sure, kindly pointed out to them.

Paul went on to the directorship of the Marine Biological Laboratory (MBL) at Woods Hole in the 1980s. Later, in retirement, he has become a prominent defender of science. He was co-author of two books. The first was with Norman Levitt, *Higher Superstition: The Academic Left and Its Quarrels with Science,* which dealt with the nuttiness of deconstructionists' claims that reality was a social construct and thus that scientific descriptions had no more claim to accurately describe reality than any others. This was the age of comments like "DNA is a construct of the male ego" or the equivalent. His second book, *Creationism's Trojan Horse,* was on intelligent design creationism and co-authored with Louisiana philosopher Barbara Forrest. Their book played an important role in understanding the wedge strategy of intelligent design and in the debunking of it.

During that time, when Beth and I were still postdocs, we visited my grown and settled cousins at their Quebec summer cottage. They pointedly asked, "Aren't you ever going to get a real job?" Yes, eventually, I had to confess, because the postdoc money would run out after two years. It was time to enter the promised-land of publish or perish. While I was at MIT, I started applying for faculty positions. There were a heady number of job possibilities advertised because the new developmental biology was riding a wave of excitement. I had several interviews at a diversity of universities in an assortment of places, which allowed me to fly on a lot of obscure and uncomfortable commuter airlines into remote college campuses. I didn't apply for any jobs in Boston or other large cities, because we had our daughter Amanda, and the city environment didn't seem like the right place to bring up a child. Our experience was that although Watertown was a nice neighborhood, it was hardly immune to city crime. There had been an armed robbery at the convenience store one block from us. This heist turned out to be a minor classic of dark comedy when the store clerk blinded the robber with mace and then dashed out from behind the counter to punch him. Fortunately, this hapless robber's gun misfired, not once but twice – first when pointed at the zealous shopkeeper and then in the face of a police officer who showed bad timing in pulling up to the curb seconds later. In accord with the irony of life rule, Amanda would grow up wanting to live in an eastern city and would eventually marry a New Yorker, Phil Cohen, a versatile movie cameraman and airline pilot, and move there happily.

My travels taught me much about the mysteries of how academic life worked. I sent one of my many job applications to the Department of Zoology at Indiana University for a position as an assistant professor of zoology to work in developmental biology. I heard nothing back from them. The chairman of the department was Robert Briggs, who was famous as the inventor of nuclear transplantation and thus revered as one of the founders of the new developmental biology. At the time, I was the postdoc representative on the seminar committee in biology at MIT. The faculty chair of the seminar committee, David Botstein, who had substantial political savvy, suggested I call Briggs and invite him to Boston to give a seminar as way to establish a contact. I nervously made the call. Briggs listened to me and happily accepted the invitation to come out and speak at MIT. Then, after a moment of silence, he said, "Didn't you just apply for a job here?" I admitted that I had, and he let drop the truth: "We lost your application. Can you send us another one?" I knew immediately that this was my kind of place.

I started at Indiana in 1971 as an assistant professor of zoology. My starkly empty lab was the site where my independent career as a scientist was to begin. The department provided enough money and borrowed equipment for me to get my lab started that first year, but I had to write a grant for substantial research funding and start recruiting graduate students in order to carry out an active research program. Things were slow at first, so while I was writing my grant and equipping the lab, I started guitar lessons. By the next year everyone seemed uncommonly pleased when I gave the guitar to Beth's sixteen-year-old cousin, Jean, for her birthday. We soon discovered that southern Indiana had clear lakes and streams for canoeing, and there was plenty of nature – hardwood forests, a rugged carst topography, including plenty of limestone caves, and spectacular Paleozoic fossils of trilobites, straight-shelled cephalopods, corals, bryozoans, crinoids, blastoids, and brachiopods. These were all the remains of life in ancient seas that had rolled over Indiana for millions of years while it lay basking on the equator. Since Indiana's seas had long ago drained away, I had to work out how to fly live sea urchins from ocean-side collectors to southern Indiana and how to keep them happy in artificial seawater. Within a few months the sea urchins made their flights from California and were happily ensconced in temperature-controlled saltwater aquaria. I was awarded my first grant – from the National Institutes of Health.

I would also take an unrelated first step into working on evolutionary biology. At this time the endosymbiotic theory of eukaryotic organisms was being propounded by Boston University biologist Lynn Margulis. She proposed that bacteria took up residence in pro-eukaryotes and were eventually evolved into cellular organelles, the mitochondria and chloroplasts. These still have highly reduced DNA genomes. I got to know Henry Mahler, who was a noted biochemist at Indiana and worked on mitochondria. We both thought that the case was not closed and that other hypotheses should be proposed. We suggested an alternate explanation to the evolution of organelles as being generated within a cell to regulate metabolism by segregating components. We wrote up our hypothesis, and it was published in *Science* in 1972. Although our account was as plausible at the time as endosymbiosis, we were proved wrong later when gene sequencing showed that organellar genomes were related to those of bacteria. No matter, our work was iconoclastic and imaginative. It also required hard thinking on an important evolutionary issue, and it gave me the confidence to return to de Beer's ideas and to start my own research on the evolution of development.

Both Beth and I would thrive at Indiana, but my career path was much easier. We started our scientific careers before the passage of Title IX of the Educational Amendments Act of 1972, which granted women equal access with men in education. When Beth first thought about applying to graduate school in the early 1960s, some graduate programs told her that they did not admit women – then a quite legal form of discrimination. Faculty positions for women in science remained scarce in the following decade and nearly impossible for women with families. When we got to Indiana, she took a postdoctoral position with Frank Putnam, a leader in the study of immunoglobulins. In 1974, as our daughter grew older, Beth pushed to be allowed to work independently and to be able to apply for her own research grants. The university responded by creating a set of not-quite-faculty-equivalent research ranks, of which she became the first holder in 1974. She applied and was awarded a research grant from the National Science Foundation. By 1984 the once-dire situation for hiring women into tenure track faculty positions had improved, and our son Aaron was now six. Beth applied for a regular faculty position that had become open in the department. She won the position and was appointed an associate professor of biology. She went on to a full professorship and eventually to serve as chair of the department from 2002 to 2007. In that time she hired twenty-one new faculty members, about a third of faculty, during the largest expansion the department ever experienced. We built separate labs and distinct research programs but collaborated on projects in the evolution of development that interested both of us.

In the years we were starting out, Beth's mother's parents lived sixty miles south of Bloomington on a modest beef cattle farm with scruffy, and barely effective, fences. Those fences made it possible for Beth to return to her childhood love of horseback riding. She bought a horse and shamed me into doing the same. My first serious riding came when we found me a suitable horse, a Tennessee walker, at a farm four miles from home, and I had to ride him uneasily back through the gathering dusk. I learned to catch, bridle, saddle, and confidently ride a horse, and to help out the blacksmith in a suitably nonchalant way on his weekend visits. I learned how to control a horse that had more fear of passing a pig by the side of the road than of an oncoming truck on the road. The intelligence of horses is only a movie legend. This was a year of learning an intense new activity, one that produced painful leg muscles for slackers who didn't ride at least once a week. Unfortunately, my adventure of riding around the southern Indiana countryside came to an abrupt end in a wild midwestern thunderstorm. The Boy Scouts have it right when they say not to take shelter from a storm under a tree. My horse was killed under a big oak tree that was split vertically by a lightning strike. His half-ton carcass was flung into the adjacent farm pond, and because there was no way to move him, there he stayed until his skeleton disarticulated and his skull became a roost for sunning turtles. Shortly after, our son was born, and our riding days faded into the sunset.

The distinct personas of our children illustrated to us the effectiveness of genetic recombination in generating variation among siblings. They were distinct people from the first moment they could express their personalities. We wondered what they would become as they grew up. We found out early what Amanda would not become. One evening Beth was reading at the dining room table where she worked at home. Amanda, then about ten stopped at her side and announced "Mom, I'm not going to be a scientist." "Ok. Why not?" "You keep terrible hours and you don't dress well." Amanda became a physician. She got it half right. She does dress well. Aaron was born rambunctious, but was logical and a natural talker. He served as a police officer on the way to becoming a lawyer and indeed became an exuberant and effective courtroom litigator.

PART TWO

Finding Evolution, Founding Evo-Devo

TO BECOME AN EVOLUTIONARY BIOLOGIST MEANS UNDERSTANDING not only the science of evolution but also the history of an evolving idea. My own entry into this science seems indirect, but studying development would let me see how major evolutionary questions could be investigated. Two great biological processes of change, the development of individuals in each generation and the evolution of organisms over eons, have together shaped the diverse body plans of living beings on Earth. Developmental genetics and molecular biology gave us the tools to delve into how development evolved and gave rise to the patterns of animal evolution seen in the phylogenetic tree. Uniting questions and tools allowed me to take early steps in the creation of the new science of evo-devo.

Your mama was a lobe-finned fish.

Ray Troll

ELEVEN

Evolution as Science

MEANING

When I was a kid and enjoyed collecting fossils and fantasizing about live dinosaurs, I was completely, even magnificently, ignorant when it came to grasping what evolution is, beyond a vague notion of one kind of dinosaur following another through time to their inevitable doom. There were many things I didn't even know that I didn't know – like what science itself is. Then there were other gaps in my knowledge, most critically of the historical origins of evolutionary biology and, most difficult to grasp, what the living science of evolution is all about. Much education happens by accident and curiosity. I have never had a formal course in evolution, and my becoming an evolutionary biologist came as a result of my reading as I became a molecular developmental biologist. I only later came to do research on the relationship between developmental biology and evolution. Now, as a whim of fate, I teach evolution to students who have no idea that I never took the course myself.

Science is about understanding the natural world. As scientists, we assume that nature operates under the actions of consistent natural forces; no miracles or other supernatural phenomena are involved. This is not to suggest that scientific reality will necessarily correspond to commonsense reality. The bizarre world of quantum mechanics makes no connection with common sense, nor can we have personal comprehension of the vast geological times we have to think about in studying evolution. Scientists assume that we can extract a consistent understanding of the natural world, and that we can test our working models, hypotheses, under tough criteria. Hypotheses should seek the least convoluted assumptions; they should not require special fudge factors; they should explain existing knowledge; and finally, they must predict the results of experimental test or new observations. Only hypotheses with a high degree of explanatory power and a history of surviving all challenges are promoted to the status of a theory. But science is always conditional. Even theories may be susceptible to being overthrown if a better model of the world becomes available. Some theories, like the rule that matter is composed of atoms, that the Earth revolves around the Sun, or that life has evolved, are strongly established and are unlikely to be incorrect.

The philosopher Karl Popper notably pointed out that you really can't prove that any of your favorite hypotheses are true, but it is possible to test by experiment or observation to see if a hypothesis is false. This may seem like an oddly counter-intuitive idea, but this principle informs us of how we should ask questions and design experiments. The critical predictive test is one that seeks to find an observation or experimental result that the hypothesis wrongly predicts. Any hypothesis that fails the test of real-world observations has to be discarded. Hypotheses that pass the test live on to be tested another day – and gain greater credibility each time they survive yet another round. In principle (if you are smart enough), you can investigate and comprehend any phenomenon in nature by devising a powerful explanatory hypothesis and applying it to guide the choice of research questions. If the hypothesis is not wrong, it will allow the prediction of the results of an experiment. If that happens, the hypothesis remains viable but is not guaranteed to be correct by that confirmation, because other tests might still prove it wrong later. The more experimental or observational tests a hypothesis passes, the more likely it is to be correct. In the case of biology, Popper once famously declared that Darwinian natural selection was untestable. Evolutionary biologists, undeterred by his comments, put natural selection to severe trials. Popper later changed his mind on this score.

It is in this sense of passing multiple potentially falsifying tests, and its broad explanatory power, that evolution is a theory. It's important to note that the scientific usage of the word "theory" does not correspond to the popular definition of the word. In everyday conversation "theory" is defined as merely an unsupported guess, as in the conspiracy "theories" of your crank uncle Harry. The resulting popular confusion over the meaning of the word has been a major debating tool of creationists. To say to a nonscientist audience that something is only a theory is an effective slam that exploits this misunderstanding of word meaning. Accordingly, if evolution is only a theory, it's just a wild idea like Uncle Harry's and hardly needs to be taken seriously. Evolution has survived over a century of stringent testing and is very much a theory in the sense used by science.

Scientists accept that the natural world exists as a reality outside of ourselves. But there is a rub – just what is it that exists? The artist Ray Troll is a devotee of both fish and evolution. Troll is best known for his paintings created around these themes, but he has composed a marvelous song, "Devonian Blues," which sings of the evolution of feet from fins, the evolution of fish to man. My favorite line in the song is "The truth is told in fossilized bone." That's absolutely right, but in fact what that truth is cannot be determined so easily merely by inspection of that fossilized bone. What does such an object tell by itself? Not much. It takes context. Interpreted with enough knowledge of fossils and biology, fossilized bone can re-create lost worlds. The fossil skulls and skeletons of ancient lobe-fin fishes tell us that our ancestors once had fish gills as well as lungs, and that feet and hands evolved from fins. Taken together with knowledge from another discipline, developmental biology, we know that all living vertebrates, including ourselves, not only had ancestors with gills, but that each of us developed "gill arches" during our lives as embryos. It's not that we had gills twice in our histories, once in evolution and once in development. We were not tiny fishes in the womb, because we don't inherit ancestral structures. We inherit genes that control the process of development of body structures. These genes and what they do are modified during the course of evolution. Human "gill arches" no longer develop gills but still produce the arteries and skeleton of the neck once associated with gills in our lobe-fin ancestors. These structures have new functions in us now that the gills are gone.

The earliest students of fossils interpreted fossils of shells or bones as having been formed right where they are found, within the solid rock, by the action of natural creative forces that they thought could mimic living forms inside of stone. Before trading science for salvation and going off to become a bishop, Nicolas Steno wrote a treatise that was instrumental in the founding of paleontology as a science in the seventeenth century. Steno demonstrated that fossils were the remains of once-living beings by a point by-point comparison of the teeth of a living shark to giant fossil shark's teeth – then called "tongue stones." He worked out how organic remains could be embedded in the sediments that first buried the remains and then solidified as rock. Finally, Steno provided one of the basic laws of geology, the principle of superposition, to explain the deposition of rock strata – older rocks on the bottom, younger rock layers above, just like the papers and unopened letters that day by day accumulate on a professor's pile of things to be answered "tomorrow."

The important clues that first pointed science in the direction of an evolutionary explanation for the diversity of life came not from biology but from a growing understanding of the Earth and its age. The collapse of the biblical timescale for creation began long before Darwin with the work of pioneering Scottish geologist James Hutton in the eighteenth century. Hutton realized that the rock record showed an immensely long history of events on Earth: mountain building followed by the mountains' slow destruction by erosion. Over time these events were followed by deposition of sediments where the old mountains once stood and the eventual uplift of new mountains. The cycle took place over and over again. The vast expanse of time it took for each event to occur could not be fit into anything like a six-thousand-year timetable. Hutton's vision was that "we find no vestige of a beginning, no prospect of an end."

After Hutton, a series of nineteenth-century geologists would become conscious that the fossil record failed to support the biblical timescale, either of the biblical sequences of creation recorded in Genesis or of the great flood. The fossil record stemmed from many distinct events of life, death, and burial over an immensely long period of time that had to break free of the biblical limit. As fossils became better understood, it became clear that the deeper in time one looked, the less like living animals the fossils were. Evolution would provide the explanation for the link between living forms and their similar but not identical precursors. It would not be simple. I've given only a tiny gloss of the brilliant intellectual history of how the science of geology was born and the struggle to date the ages of the Earth. I've noted two books by Pascal Richet and by Martin Rudwick in the bibliography for this chapter that do a lavish job of it.

Geologists in the twentieth century refined the concept of geological deep time and developed the radioactive decay dating methods, based on unstable elements like uranium, which established a history for the Earth that extended back over billions of years. Radioactive decay dating depends on determining the amount of decay of radioactive isotopes that have taken place since the original formation of a rock. The tools and calibrations for radioactive decay dating came from nuclear physics. Creationists have raised frivolous and unsupported objections to dating, such as the suggestion radioactive decay rates vary over time, but physicists haven't found the longed-for variation. Sadly for the die-hard supporters of the notion of a six-thousand-year-old world, if nuclear submarines work, then radioactive decay dating does too. It relies on the same physical principles, and its application of these principles to both the rocks of the Earth and to meteors always says "real old." Hutton was mistaken in one way, though. The modern science of radioactive decay dating has allowed us to find a vestige of the beginning of our Earth. It happened 4.6 billion years ago. Astrophysics allows us to also predict an end. We are about half way through the life of the Sun.

The new science of geology replaced the metaphorical biblical flood story with a more realistic and longer-term history of events. Fossils gave us stony evidence of a changing series of life-forms representing the history of living creatures on Earth. The idea that fossils directly provide us with evolutionary information is more recent. If the fossilized bone of "Devonian Blues" had been found in medieval times, it might have been placed in a church reliquary as the relic of a saint. Fossil remains found in the period following Steno were widely interpreted as remains of the victims left by Noah's flood. Only by the time of Hutton and Darwin was it possible to see them as ancient animal remains, unrelated to any biblical stories, telling us something about the evolution of past life and of the time in Earth history when they lived. The point is that a natural "fact" (a fossilized bone that you hold in your hand) may be true, but what it or any other fact means is not self-evident without other data and ideas. A body of interpretation based on testable theoretical concepts is needed.

Biology is firmly a part of the natural world and obeys the laws of physics and chemistry, but life does things that go beyond anything seen in nonliving nature. Life sustains itself, using the thermodynamics of open systems to take in energy and build order. That is, energy is taken in by the organism from outside itself and supports the formation of an ordered and dynamic structure. When that system fails, an organism dies and its components are reduced to an equilibrium state. Its complex components decay to stable (sometimes smelly) simpler ingredients. Living beings can reproduce copies of themselves through a process that might seem like complexity arising from the formless mass of an egg. The formlessness is an illusion. Cells, including eggs, contain order in the form of the genetic instructions encoded in DNA. A century of genetics has shown that the encoded information and a translation system that expresses that information in cellular components underlies and informs the development of a new individual. The simplest form of reproduction is through copying this encoded information. The original cell divides into two daughter cells, and an identical copy of the genetic information goes into each daughter cell. This is what single-celled organisms such as bacteria do. Multicellular organisms have special reproductive cells (eggs and sperm) for the recombination of information and the generation of new cells in the growing individual. Shuffling of genetic information takes place, and so do errors. In consequence, new individuals are never identical to their parents. Life begets complexity, and because errors creep into the information encoded in the DNA, life also begets variation. That is one of the crucial founding principles that drive our understanding of evolution.

Darwin discovered two of the major themes in evolution, common descent and natural selection. He intuited that descent with diversification was the inevitable consequence of the processes of natural selection. Darwin's 1859 arguments for how evolution works actually still encompass the process well. Thus hereditary information has to be passed on from parent to offspring. But sometimes there is a change, what we now call a mutation, and that produces variation. Individuals thus will differ in the genetic variants they possess. As the number of descendants that survive is limited, if a variant favors the survival and reproduction of an individual over another, that variant, if heritable, is passed on and will increase in the population. This is natural selection. A sliver of evolution will have taken place. This process summed over the enormity of time has produced all life on Earth from our long ago common ancestor, a primitive single-celled organism.

CONTINGENCY

Life has a history. It had an origin, and it evolved its current diversity on Earth by descent with modification from that single common ancestor. Skunk and skunk cabbage, not to mention bacteria, yeast, lettuce, mushrooms, spiders, apes, and people, are all related by common descent – an extraordinary but inescapable conclusion. The components of a living being obey the laws of chemistry and physics. Yet the path of future evolution is not predictable. The complex association of features produces emergent properties, that is organisms and behaviors that cannot be predicted from the underlying hierarchy of rules. Some emergent properties can be seen at nonliving levels. Hydrogen and oxygen are gasses. Combined, they form water, which is wet and a liquid at room temperature because of polar attractions of water molecules to each other. Water ice is less dense than liquid water and so floats. These properties happen to be vital to all life as we know it. Among living organisms, gorgeously unpredictable emergent features arise, and evolution itself is emergent from the properties of the replication of information and competition between individuals. The most notable emergent features to us are those of mind and behavior. Nothing in physics requires the existence of bird's-nests or table manners. Do we imagine that music and math abilities were directly selected for? More likely they are byproducts of the abilities of a wonderfully intricate mental machine evolved under selection for other functions and indeed, some other organisms, such as our close relatives the chimps, also show emergent mental features like self-awareness and humor. Show a chimp a mirror. He gets that he is seeing his own reflection.

The second factor that makes the history of evolution unpredictable is its contingency. Things happen by accident. This is the feature of evolution that is hardest for many people to bear, because of their desire for a purpose-driven origin for each and every one of us. The emergence of human intelligence by evolution inspired the French Jesuit and paleontologist Teilhard de Chardin to propose in his influential book *The Phenomenon of Man* (a work composed of dense language, a thicket of neologisms and some spectacularly convoluted writing) that there was a "process of humanization." He regarded human evolution as directed at reaching a higher goal, a universal intelligence. If we look back at events that have already occurred they look inevitable. Are they? My own existence depended either on a guided meeting of two people born on different continents or on their chance meeting. If they hadn't met, some other outcome would have looked inevitable, but it wouldn't be me writing about it.

Like human history, evolutionary events and evolutionary history result from a mix of pre-existing conditions and unpredictable accidents. There was no guarantee that dinosaurs would be wiped out by that asteroid at the end of the Cretaceous, unless you believe that asteroid was guided to Earth with the purpose of producing that finale. But if it were, why did the owner of the guiding hand wait those nearly 200 million years of the Mesozoic to get rid of the expendable monsters? Why not just get on with the main show – people, intelligence, self-awareness – much earlier? Perhaps if dinosaurs had not been so decimated, they instead of mammals would have given rise to an intelligent and conscious species. Some of their descendants, the larger parrots, show some language ability and a liking for dancing to music, so the evolutionary potential for intelligence may well have been incubating in their dinosaur forebears. Whether the evolution of intelligence and consciousness is inevitable once sufficiently complex creatures evolve is an engrossing question that will be hard to settle without the discovery of other planets with life. The important thing to realize is that emergence doesn't trump contingency. Emergent properties arise naturally given the right conditions, but those conditions are dependent on contingencies. In mass extinctions, well-adapted organisms can be driven to extinction, completely resetting the evolutionary stage in unpredictable ways. Earth history has seen plenty of that.

Two scientists of the sixteenth and seventeenth centuries, Copernicus and Galileo, removed our planet from the center of the solar system and eventually from the heart of the universe as well. That was jarring, but it was just the first blow science would deliver to the human ego. Darwin's theory of evolution in the nineteenth century would remove humanity from its dream of being the centerpiece, the meaning, and the very purpose of the creation. In fact, physics, geology, and evolutionary biology together would obliterate the creation itself. Darwin's key concepts still drive much research in biology, and biology is hardly comprehensible without the concept of evolution at its core. Darwin's contributions in the *Origin* were epochal. His work revolutionized biology, but more fundamentally, it changed our view of the world and the place of humanity in it, something still ferociously denied by many religious people. Few scientists have the following that Darwin does. He was an appealing and accessible historical figure. *The Voyage of the Beagle,* his letters, and his *Autobiography,* as well as innumerable biographies over the century after his death, have ensured that he remains a completely human persona to us. Yet Darwin's connection to the modern study of evolution needs to be considered, because although he saw much, he didn't see everything.

Charles Darwin covered in every Bio course I've taken –
just a little redundant.

Anonymous, student course evaluation comment

TWELVE

Dining with Darwin

THE RELUCTANT TOURIST

Biologists like to have a sense of connection with places or events associated with Darwin. That comes mainly from reading *Voyage of the Beagle,* which along with the great Darwin biography industry has given us an amazing sense of intimacy with him, a sort of feeling of kinship. I feel it, too, and I have enjoyed my contacts with Darwin's traces even though I've never tried to follow his *Beagle* travels. The first Darwin site I saw happened to be the place where he spent most of his post-*Beagle* life and where he wrote *Origin of Species.* That was Down House in Kent, about sixteen miles south of London, which Beth and I visited in September 1974. Our trip there was long before Darwin's popularity peaked again. Down House was empty of visitors, so we could look around the garden in solitude and stroll the famous sandwalk where Darwin walked for exercise and to think. As we were alone in the place, the custodian opened the cord blocking casual entry into Darwin's study and showed us around, a treat that may be harder to come by now. The room was preserved pretty much as it had been in Darwin's time. We stood in front of his desk, his mantel, and his bookcases. I put my hand on his desk. This was the room he wrote in. You might imagine that you could speak to him there, but you can't reach across time except in imagination.

Darwin published the *Origin of Species* in 1859. The book led to immediate controversy. The first major scientific skirmish over it came seven months later, in the legendary Oxford debate of 1860, which made Oxford another site of importance to the history of evolution, despite the fact that Darwin himself was not present. Although there was no transcript kept of the meeting, the event has often been described as a clash between Bishop Wilberforce and two of Darwin's prominent supporters, botanist Joseph Hooker and morphologist Thomas Huxley. It is there that Huxley famously responded to a gibe by Wilberforce who asked him if "it was through his grandfather or his grandmother that he claimed descent from a monkey." Huxley is reported to have responded, "If then the question is put to me whether I would rather have a miserable ape for a grandfather or a man highly endowed by nature and possessed of great means of influence and yet employs these faculties and that influence for the mere purpose of introducing ridicule into a grave scientific discussion, I unhesitatingly affirm my preference for the ape." Women were supposed to have fainted at Huxley's rude comment to the worthy bishop, but I think basically most attendees must have had a jolly time of it. Huxley, for his part, was promoted to the status of "Darwin's bulldog." Robert FitzRoy, the former captain of the *Beagle,* stood up to thrust his Bible up over his head and urge people to heed the word of God. Darwin's captain and shipmate never forgave him.

The study of ancient life had come to Oxford before Darwin in the person of Professor William Buckland. The first fossils ever collected of dinosaurs are at Oxford, including those of the giant meat eater *Megalosaurus,* the first dinosaur ever described – by Buckland. These first dinosaurs were known only from incomplete remains and thus were fancifully restored as elephant-sized lizards by the artist Waterhouse Hawkins for the 1851 Great Exhibition (essentially the first world's fair). His giant replicas, carefully arranged in geological order, still bask in a sunlit park in the London suburb of Sydenham, at the site of the Crystal Palace, which housed the fair. That structure has long since burned down, but I can attest that the monsters still haunt the park. The *Iguanodon* model is so large that Hawkins once hosted a banquet for scientists in the shell of the half completed body. Dinosaurs played a role in Richard Owen's opposition to Darwin. Owen noted that dinosaurs were much more highly developed than living reptiles. He argued that if dinosaurs were as advanced as modern mammals, they were hardly consistent with a progressive evolution from that era to the modern world.

Oxford also owns William Smith's original, giant hand-drawn and color-coded map of the geology of Great Britain. The map had been in the possession of Smith's nephew, John Phillips, Darwin's contemporary and the Oxford professor of geology who named the Paleozoic, Mesozoic, and Cenozoic eras of life on Earth, terms still used today. Smith had drawn the world's first-ever geological map of a country by tracing geological strata across England and showing their ages by use of different colors. Smith's map was a revolutionary thing in 1799. It was an act of genius by a canal engineer of low social status. His revolutionary contribution was that he recognized that fossils were the key to unlocking the relative ages of rock strata, and that strata of the same age could thus be mapped over great distances. Because the science of the time was an occupation for wealthy gentlemen, it was late in life before Smith received the scientific recognition due him.

I was in Oxford in 2004 to give the Jenkinson Lecture as the guest of my friend Peter Holland, the Linacre Professor of Zoology. The lecture honors a brilliant young anatomist tragically killed at Gallipoli in World War I. Peter arranged a visit for me with James Kennedy, a paleontologist and the director of the Oxford Museum of Natural History. He was a wonderful guide, knowledgeable of history and patient, who showed me around. I admired the Buckland touch, which was apparent in the display of his collection of coprolites, the nice name for the fossilized turds of giant marine reptiles and sharks. His coffee table, made from a tastefully arranged set of cut and polished cross sections of coprolites inset into the top, is even now on display. The Oxford debate had taken place in a hall in Oxford's Natural History Museum. A famous and cherished room, surely, so I asked if I could see it. Jim said yes but warned me I might be disappointed. The entry into the room was an obscure door opening off of the main exhibit hall. The reason I might be disenchanted soon became evident. The meeting hall had at the time of the debate been a tall and quite narrow room. This lofty space was not preserved in its original glory. As the excitement of the 1860 controversy faded, it had been demoted to mundane museum use, a storeroom for scientific specimens. The original hall was tall enough that a floor had been added midway up to create a second level for yet more storage. Jim took me upstairs to see some of the original architectural features not obscured by glass-topped insect specimen cases. Much of that second floor was devoted to drawers of butterflies, including some of those collected by Alfred Russell Wallace, and even some collected by Darwin. It was still an evocative place.

We ran across Darwin in Australia in more understated ways. Beth and I spend our Decembers doing research in the School of Biological Sciences at the University of Sydney, located on Science Road. When Darwin passed by on his visit to Parramatta, the farm property upon which the University of Sydney would be built lay next to the main road from Sydney town to the New South Wales administrative center of Parramatta. The first construction of the university, the Oxford University–like main Quadrangle, was built of the local streaked brown Hawkesbury sandstone in 1850. The gaggle of gargoyles carved into this building would do any medieval European cathedral proud. Darwin had no relationship to the then non-existent university, but the university owns one curiously mundane artifact of Darwin's travels. This is a chunky four-sided brown sandstone milepost that once stood at the edge of the old Parramatta Road. It stands on Science Road, partly concealed by foliage just outside of the building we work in during our stays in Sydney. The milestone was carved in 1810 and marks a distance two miles from Sydney and thirteen miles from colonial Parramatta. The rumble of the heavily trafficked Parramatta Road lines the northern boundary of the modern campus. Darwin would have ridden past the old mile marker on his ride to Parramatta in 1836, but contrary to the campus legend I was told, he did not trot down Science Road. The stone was removed to Science Road following development of the science precinct in the 1860s.

After spending a few days in Sydney, Darwin determined to visit the Blue Mountains and socialize a bit with an English gentleman living in Parramatta, Captain Philip Parker King of the Royal Navy, who had retired from a career of charting the Australian coast and moved inland to build a large estate in the quiet backcountry. There were close ties between King and the *Beagle.* Captain King had commanded the naval squadron that included the *Beagle* on its previous surveying voyage to South America. When the *Beagle's* captain, Pringle Stokes, committed suicide during a bout of severe depression, FitzRoy was appointed to replace him. King was the father of one of Darwin's shipmates on the second voyage of the *Beagle*, midshipman Philip Gridley King. King's estate would have seemed like a precious bit of the England to which Darwin longed to return.

Curiously, Darwin's remarks and notes during his time in Australia read like the comments of a sophisticated and somewhat grouchy tourist who has been forced to visit just one too many historical sites rather than the observations of the prospective father of natural selection. Darwin records little about the Australian fauna and remarked in the *Voyage of the Beagle* that he was not much impressed by New South Wales. Darwin's note on first viewing the scrub vegetation of the cliff tops around Sydney reads, "The nearly level country is covered with thin scrubby trees bespeaking the curse of sterility." One can only conclude that five years away from home was enough and that Darwin's enthusiasm was fading. Australian social development left him cold. He acknowledged that it had prospered as only a colony founded by England could, but he found the sight of wealthy former convicts riding through the streets of Sydney in their carriages offensive to his sense of propriety. Nor, to his credit, was he attracted by the continued use of convict labor. He concludes, "My opinion is such that nothing but rather sharp necessity should compel me to emigrate."

My own first view of the New South Wales coastal scrub around Sydney and its sandy soil also struck me as a poor substrate for creating agriculture. I would have felt despair if I had been told that I had to plow that ground and grow my food there. But from the perspective of a well-fed visiting biologist, the plants illustrate remarkable evolutionary origins and extraordinary adaptations to dry conditions and to fire. The *Banksia* trees, which are relatives of the *Protea* trees of South Africa, produce grotesquely large seedpods. These require being burned over by fire to open so that their seeds can germinate – that tells you something about the fierce ecology of the Australian bush.

Considering that Australia has evolved in isolation for 60 million years, it is surprising that Darwin didn't exhibit the enthusiasm one might have expected of him in entering an isolated continent full of unique creatures. In South America, Darwin discovered the connection between the unique living beings of the continent with their fossil relatives, also confined to South America. The Galapagos Islands were stunning, and their island faunas would reveal a hotbed of evolution. Australia possesses an even more impressive and older natural laboratory, but Darwin had only a short time to spend in Australia, and to be fair, he was getting weary. The Galapagos consists of a small theater of evolution consisting of a few islands and a short span of geological time, whereas Australia is a vast and ancient continent – a far more complex prospect than the more obvious and simpler evolutionary phenomena of small islands. Darwin made one famous note foreshadowing his developing views on evolution. He observed in his 1839 *Journal of Researches* that many creatures of Australia are similar to those of other continents, although not necessarily closely related. He commented, using conservative language, that such similar results arising on two continents implies something important: "Surely two distinct Creators must have been [at] work; their object however has been the same & certainly the end in each case is complete." In modern terms these creatures have evolved their similarities independently in two continents, a phenomenon called convergent evolution.

The Blue Mountains west of Sydney beckoned, and Darwin continued his inland travels to view their geology. He enjoyed the spectacular scenery of the crevasse where Jamison Creek suddenly dives off a cliff to become Wentworth Falls. There is now a Darwin Walk along the creek, which ends at the cliff edge viewpoint where he looked over into the vertical chasm. In Darwin's time there was no way to descend the cliff there, but in 1906 a spectacular (and from my experience as someone with no enthusiasm at all for heights, dizzying but wonderful) trail was built that allows one to walk down a narrow track pressed against ferny ledges on the cliffs near the waterfall. These wet areas are vertical bogs that have gorgeous carnivorous plants covering them, sundews with their red mittens showy with sparkling drops of fatal glue.

Darwin wrote a book called *Insectivorous Plants* in 1875 in which he details the prey capture structures of carnivorous plants and describes extensive experiments on them. Nearly seven hundred species of insect-eating plants have been discovered. They have evolved several times, and they have no single optimal "design." Some similar trapping systems had multiple origins; for example, pitcher traps arose independently in four different families of plants. The highly distinctive killing tools of the different groups of carnivorous plants evolved from otherwise standard plant parts. Venus flytraps have jawlike snap traps, sundews and a couple of unrelated families use glue, pitcher plants drown insects in treacherous water-filled vases evolved from leaves, and bladderworts have underwater suction traps. Darwin called these derived structures "contrivances," which unfortunately carries the notion of design. But even now, we don't really have words to refer to elegant biological structures that don't have teleological nuances. Perhaps it's like comedian Stephen Colbert's word "truthiness" for confident statements that sound like the truth but are not that at all. When we say words that imply design in the context of biological organization, we really mean "designiness," the deceptive appearance of design. The fact that we don't have such terms means that the limitations of ordinary language inhibit our accurate descriptions of organisms and how they evolve.

To refer organisms or their parts as machine-like implies design and can lead anyone listening to how we describe organisms astray as to what we are suggesting about causality. There is a history of biological interpretations being misunderstood in this manner. So to human eyes ants seem so organized that they have often been thought of as being controlled by top-down "orders" from the queen. Recent studies instead show that the overall activity of ant colonies is the sum of contingent individual responses made as a result of stimuli from contact between ants in a shifting network of contacts. There is no intelligent design or command, despite how much the operation of an ant colony looks like a command organization on the surface.

Darwin also mused about the human condition, and he was curious about the Aborigines he met in his travels. He was intensely interested in humans as they exist in the "wild" versus civilized state. He regretted the coming fate of the Aborigines as they lost their lands and cultures to the spread of the settlers. Darwin made notes on the geology of the Blue Mountains, consistent with his main interest as a geologist during the voyage, and developed an (incorrect) hypothesis as to how the great cliffs had formed from action of the sea. In contrast to his efforts in South America and his working out how Pacific coral atolls form, this wasn't one of his better geological efforts. For entertainment Darwin went riding after kangaroos and shooting platypuses, but he really did not seem to think deeply about the meaning of the weird zoology of Australia.

Perhaps most unlike the Darwin who had enthused over rheas in Patagonia was the lack in his writings of any significant mention of the platypus. Here is an animal even few Australians have seen in the wild. This is the most oxymoronic of creatures, an egg-laying mammal, and a survivor of the Mesozoic era. The living platypus reveals characteristics present in ancestral mammals a hundred million years ago, especially its weird mosaic combination of mammalian warm-bloodedness and fur harnessed to a reptilian reproductive tract. From 1800 on, the platypus was at the center of major arguments among zoologists. The premier anatomist of the time, Richard Owen, entered the fray in the 1830s without solving the puzzle. At the time Darwin visited Australia, the milk glands of the platypus had just been spotted for the first time. How platypusses reproduced was still completely unknown, although the disturbing possibility that they laid eggs was whispered. As early as 1803, Aborigines had told Europeans that the platypus laid eggs in burrows, but their stories had apparent inconsistencies because of translation difficulties. Besides, European scientists asked, what could unlettered Aborigines know in the face of anatomical experience that proper mammals simply did not lay eggs?

The eggy conundrum was not solved until 1883, when an ambitious young embryologist named William Caldwell came out from Cambridge University on a mission to investigate two potential living fossils to see what their development would tell about evolution. His first target was to find out how embryos of the Australian lungfish, *Neoceratodus,* developed. Lungfish are related to the lobe-fin ancestors of the evolutionary radiation of vertebrate animals onto the land. It took considerable labor and help from knowledgeable Aboriginal people to find fertile lungfish eggs. Caldwell demonstrated with his work on the lungfish that he had no desire to fill in minor footnotes of bits of life-history lore. He then moved onto a military-scale pursuit of echidnas and platypuses. Caldwell and a team of 150 Aborigines tracked them down during the breeding season and discovered that both produced eggs. The cost was high in numbers of echidnas and platypuses sacrificed to science, but in 1884, Caldwell published his demonstration that these mammals produced eggs like those of reptiles. Caldwell was working in the heroic height of the period of Haeckel's approach to the tree of life. This was the powerful idea that embryos would tell all about the history of evolution. What could have been more revealing than the development of such spectacular transitional forms linking cold-blooded reptiles with warm, furry mammals? This was a triumph of Darwin in a nineteenth-century setting.

EVOLUTION AFTER DARWIN

Over time, Darwin has been elevated to a larger than life figure by evolutionists, which has had one unfortunate influence on the public understanding of evolution. I think people come away from our celebration of Darwin's seminal importance with the mistaken idea that Darwin equals evolution. We biologists know that's not so, but imagine what it looks like to an outsider who might interpret our interest in Darwin as a worship of the bearded prophet of a cult of nature, and what seems to be an unshakable belief in his unchanging teachings. Thus the field of evolutionary biology can be portrayed as an ossified belief system completely frozen in the year 1859 with Darwin as its deified father figure. This viewpoint is beloved of creationists. It means to them that evolution too is a religion. It means that if Darwin was mistaken on any point or didn't know about something, the entire "dogma" of evolution is wrong. A reflection of this notion is unfortunately frequently trotted out in science reporting, in which some result is advertised in a headline that claims it has "overthrown Darwin's theory" or the like. It is shopworn journalistic hype to dramatize the writer's story, but it works to make headlines. I offer a sampling of "X Discovery Challenges Darwin" headlines from many examples of this journalistic formula that I found Googling one afternoon when I should have been working on something else:

> "Lake Louise Has Fossil Beds that Challenge Darwin"
> *New York Beacon,* 2005 (topic: Burgess Shale fossils)

> "Why Darwin Was Wrong about the Tree of Life"
> *New Scientist,* 2009 (topic: molecular biology)

> "Biologists Rethink Darwin's Theories on Sex Roles"
> *Women's News,* 2003 (topic: evolutionary theory)

"What Darwin Got Wrong about Human Emotions"
Peak Oil Blog, 2010 (topic: sex and reproduction)

"Rare Pink Iguana Eluded Darwin and Others"
Live Science, 2009 (topic: zoology)

"Swedish Chickens Challenge Darwin"
Local, 2009 (topic: animal behavior)

None of these funny titles really mean much, not even those concerning the iconoclastic Swedish chickens or the elusive pink iguanas. If Darwin had been mistaken about or hadn't heard of something hyped in one of those headlines, that fact likely has little impact on the workings of the modern science of evolution. Evolutionary biology is a vibrant and active research discipline that has incorporated Darwin's insights into a vastly enlarged understanding of biology and geology built on discoveries not available at his time. Evolutionary biology encompasses many entirely new disciplines that arose long after the appearance of *Origin*. The modern disciplines of plate tectonics, microbiology, genetics, ecology, developmental biology, molecular biology, genomics, and informatics that are now contributors to evolutionary studies were all created long since Darwin paced the sandwalk behind his garden and thought out his ideas. On one hand, the new disciplines and data have given us completely novel hypotheses and information that have revolutionized science. On the other, the new results have enlarged and enriched Darwin's concepts without displacing them. The new data keep strengthening our understanding of evolution. We have now overwhelming evidence in support of the fact that life has evolved, that all life on Earth has a common ancestor, and that natural selection works. Evolution is a lively and argumentative field, a living science.

Darwin's key arguments were about a theory for an evolutionary mechanism, natural selection. Common descent and the evolution of life were widely accepted from Darwin's time, but natural selection was doubted or rejected by leading scientists by the end of the nineteenth century, and other mechanisms of evolution were suggested. This situation only turned around in the twentieth century, when mathematically expert geneticists developed ways to account for population-level genetic phenomena, including selection, mutation, and migration of genetic variation among populations. This body of work became known as population genetics and it provided a theoretical foundation for the study of selection.

At first, the demonstration of selection in nature was difficult to do. One of the most famous examples from the pioneering period was the phenomenon of industrial melanism in the British peppered moth. The moth is historically a speckled gray color, but with the darkening of tree bark in the industrial midlands of England during the nineteenth century by coal soot, the fraction of speckled moths in the region declined and a black variant became prevalent. In the mid-twentieth century, English biologist Bernard Kettlewell carried out an enormous set of experiments to test the hypothesis that moth coloration was under selection by the feeding of birds on moths. The idea is simple: The birds ate the individual moths they could see more easily sitting on tree bark. Gray moths are harder to see on gray bark – vice versa if the bark is black. His studies were hard to do, but they provided the first major demonstration of selection in nature. Kettelwell's work was later challenged on grounds of the methods he used to test predation. Creationists avidly seized on the controversy as an indication of fraud and, in their usual subtle fashion, generalized this controversy about one study into a general denial that natural selection occurs. Selection between color forms of these moths has been re-investigated with much better methodology in the past few years, which confirms that selection on color morphs by hungry birds does indeed occur. Other well-designed and clever experiments in nature with bacteria, fish, and birds have further demonstrated that natural selection operates in nature wherever life pokes up its head.

Darwin's second major contribution was the concept of common descent. I've wondered why Darwin did not illustrate *Origin of Species* with pictures, for example, illustrations of varieties of pigeons or diagrams of artificial selection. He used extensive pictorial illustrations in his other books, and other people have gone so far as to produce illustrated editions of *Origin*. It seems that Darwin was publishing under pressure. Alfred Russel Wallace, a young biologist working alone in the islands of Indonesia, had sent him his independent conception of the role of natural selection. Wallace and Darwin published back-to-back short papers announcing the discovery of selection in 1858. Darwin had been laboring for nearly two decades and knew that now his priority could be lost. After years of dawdling, he was now in a hurry to get his book out. This hurry, and his intentionally structuring the book as a well-crafted verbal argument to sell what would be an unpopular idea, would have mitigated the need for extensive illustrations. *Origin* appeared in 1859 and sold out on its first day.

Darwin did use one illustration in the book, but it's not about selection. It is about another completely novel idea, common descent. The diagram is of a hypothetical phylogenetic tree – the first ever published. Phylogenetic trees portray evolutionary family histories. Essentially, a phylogenetic tree shows an ancestral form and its descendants. Thus the first species gives rise to two others. In turn, these give rise to other descendant species or even become extinct – a dead limb on the branching, tree-like diagram that results. It is this aspect of Darwin's work that addresses the appearance of the history of life, as we should expect it in the fossil record. Darwin saw that two conclusions were inevitable. First, all life on Earth had a common ancestor. Second, there is a branching pattern to relationships in the biological world because of divergence in form as species separate. Darwin spent less time on this aspect of his theory than he did on selection. The reasons seem clear. Natural selection was the mechanistic linchpin for all else, including the resulting pattern of phylogeny. Thus selection had to be defended before any other issues could be taken up. But that isn't all. Darwin's discussions of the fossil record in *Origin* show how disturbed he was by how much it seemed to lack in evolutionary connections. Speciation by descendants gradually diverging from a common ancestor and one another should leave behind a record of many transitional forms, but in his time, the fossil record was largely silent. He thus spent a fair amount of time apologizing for the gaps and flaws in the incomplete record revealed by the paleontology of his day.

Darwin's worries about the completeness of the fossil record have been largely resolved by the far more extensive fossil graveyards discovered since his time, but we are also aware that the fossil record can never reveal the entire history of life. Many creatures were never preserved because most organisms are small and have no skeletons. Some soft-bodied organisms have been fossilized, but they represent only a tiny fraction of the diversity of soft-bodied forms at any time. In modern shallow marine environments about 70 percent of the animals are soft bodied. Organisms that live in environments, such as mountains, where erosion dominates will never be fossilized. Others will never appear because their populations were small or they were evolving rapidly. Yet others were fossilized but the rocks containing their remains have eroded away – whole species turned to dust. As we go back in geological time, more rock is lost, so that there is a bias against older and older fossils. Thus we know that there are systematic biases in the fossil record and that they work against us. Fortunately, paleontology itself is an active discipline that aggressively seeks evolutionary sequences of descent. It has developed sophisticated statistical methods of dealing with the problems arising from the rarity with which living individuals become fossils, and it has developed improved ways to study the structures of fossils and to interpret them. Finally, since Darwin's time an increasing area of the Earth's surface is being explored for fossils, including huge fossil-rich China. The more rock examined with an expert eye, the better the chances of turning up the rare and unusual. These discoveries have begun to provide the unambiguous examples of intermediate forms that Darwin lacked.

The first spectacular intermediate between birds and reptiles was discovered in 1860, only a year after the publication of *Origin*. This was the most famous fossil find ever, the first fossil of an ancestral bird, the full skeleton of *Archaeopteryx* discovered in southern Germany from rocks of the age of the great Jurassic dinosaurs. This ancient species had birdlike feathers preserved in place in the fossil, and it had wings – but it also had the skeleton and teeth of a small carnivorous dinosaur. Here was an unexpected link between distinct classes, an astonishing transitional fossil between bird and carnivorous dinosaur. Remember that dinosaurs were thought of as cold-blooded, scale-covered, and sluggish reptiles, like big lizards. Birds are hot-blooded, feathered, and intelligent. Daring to link them in evolution was astounding.

Archaeopteryx was followed a few years later by a menagerie of fossils that documented much of the evolution of horses from browsing tapir-like creatures with many toes each bearing small hoofs to the present-day grazing horses with just one large hoofed toe. In our times, during the last few decades we have seen even more astonishing fossil discoveries. We now have a record of fossil bacteria and microbial structures that document life from the deepest history of the ancient Earth, and we have the spectacularly preserved remains of the first complex animals. In addition, there is now an informative bony record of the evolution of land vertebrates from fish through many intermediates – essentially from fish with fins to fish with legs and toes. Fossil intermediates now even show us such bizarre histories as the origins of flatfish (like flounders, the ones with the funny eye that migrates across the midline of the fish during development so the adult ends up with both eyes on one side of its head). Their ancestors were ordinary bilateral fish with an eye on each side of the head, and the intermediates have an eye partially migrated.

The seemingly improbable origin of whales from land mammals is now known from a revealing sequence of fossil skeletons of stages from land mammal through semi-aquatic intermediates to fully aquatic primitive whales. Other fossils show the evolution of snakes from lizards with legs via ancestral snakes that still possess well-formed hind limbs to advanced snakes without legs. Paleontologists have discovered a profusion of new fossils from China that preserve in detail the entire bodies of feathered dinosaurs and illustrate much of bird evolution as well as the stages of evolution of feathers themselves. These fossils also show that the dinosaur cousins of birds were feathered and almost certainly warm-blooded and highly active. In some, pigment cells have been fossilized, which has allowed inferences to be made of the colors and patterns of the plumage of small, feathered dinosaurs. A feathered *T. rex* – it seems ludicrous, but it's likely.

Human ancestors are part of this assemblage of transitional forms, too. Our human fossil record is remarkably good, with fossils showing the evolution of early hominids from ancient apes, and then of modern humans from more advanced hominids. We have a growing tally of well-preserved and even spectacular fossils showing the evolution of our own species from our first steps of upright walking. The year 2009 saw the publication of our earliest well-studied deep ancestor, the 4.5-million year-old *Ardipithicus ramidus.* This creature has a hip intermediate in structure between that of apes and upright walking humans. Its foot shows that its big toe was thumb-like and capable of grasping branches when it climbed. *Ardipithicus* provides intermediate features unimagined even a few years ago. There is an extensive 4.5-million-year record of human brain evolution showing an increase in brain volume in progressively younger members of our lineage from the ape-sized brains of our earliest ancestors.

Thus the fossil record has been transformed by discoveries over a century and a half from being the liability that had so worried Darwin to a highly informative record of new organisms arising through transformations of ancestors and descendants in the course of evolution. These fossils reveal evolutionary transitions and fill in lost branches in the phylogenetic tree of life with enlightening creatures at many levels of evolution into the deep past of the planet. Intermediate forms abound for those willing to see them.

The greatest missing element to Darwin's theory of evolution was that he could not account for how heredity worked. He had no idea what variation was, how variants could be transmitted without being diluted out in the offspring, or how variation arose in the first place. This was a weak point that wasn't missed by his eager critics. The brilliant polymath engineer Fleeming Jenkin in a review written in 1867 pointed out that the idea that artificial selection was a model for natural selection was not persuasive. Both Darwin and Jenkin believed that heredity was blending, which would mean that a rare favorable variation could never spread in a population but would be diluted out like a drop of ink in a lake – a fatal outcome if true.

Ironically, the question had been answered, but neither of them knew of the work of the monk Gregor Mendel of the Augustinian abbey at Brno in Austria-Hungary (now in the Czech Republic). In 1865 and 1866, Mendel published his experiments showing that inheritance was not blending but particulate. No one noticed. Despite the legend that Mendel sent Darwin his papers, there is no evidence that he did. Darwin did attempt to formulate his own theory of inheritance, pangenesis, in which particles, which he called "gemules" were released from each organ, traveled through the blood, and entered the reproductive organs. Pangenesis was not a success and was quickly disproved by Darwin's cousin Francis Galton, who made transfusions of blood from black rabbits into female white rabbits. Their offspring when mated to white rabbits should have shown black markings derived from the particles introduced from the blood of black rabbits. They didn't. The inheritance of acquired characters theory died at the hands of other experiments as well. Cutting the tails off of successive generations of mice did not produce offspring with shorter tails. Mendel's work was ignored until it was rediscovered independently in 1900 by three separate research groups. It was soon realized that Darwin's problems with heredity could be solved by Mendelian genetics. Population genetics would make that link.

Today's evolutionary biology has a completely novel understanding of the phylogenetic tree concept that comes from a source of data not dreamed of in Darwin's day, the opulent history of life recorded in the genes of every organism. Darwin's ideas came from the data of comparative anatomy. We have a new anatomy, that of the DNA molecules themselves. Genes are encoded in a giant molecule, a strand of DNA, which can be sequenced to gain the order of the nucleotide code words along the strand, like the letters of words in a sentence. Within the past two decades, methods to directly determine the DNA sequences of organisms have yielded a new independent set of data not used in the original reconstructions of descent. The enormous amounts of evolutionary data that can be gathered from genes are revolutionary. Sequences of genes from different species can be compared to see how they change with relationship. The more distant in evolution two creatures are, the less alike their gene sequences are, because after two species diverge, their genes accumulate more and more distinct changes. The sequences of genes allow the building of phylogenetic trees independent of trees inferred from anatomy. Astonishingly, gene trees confirm the general outline of relationships derived from comparative anatomy and the fossil record, providing crucial independent confirmation of evolutionary descent. This ability to read the history of life from genes was a remarkable result, because Darwin's hypothesis of common descent was completely innocent of the concept of DNA or genes.

The branching form of the phylogenetic tree at the gene level is consistent with Darwinian evolution. The first key point is that DNA sequences can be conservative over distantly related organisms. This means that crucial DNA sequences that are under strong selection because they perform critical functions would change slowly, retaining identifiable related sequences even after enormous lengths of time. This feature of DNA evolution lets us identify genes shared among all the kingdoms of life. Less generally crucial sequences would change more rapidly and let us see evolution in more closely related species. Second, after populations bearing the ancestral genome separated, differences in DNA would have started to accumulate and even be differently selected upon in the descendant populations. Thus divergence in genes occurred through time. Gene sequence data went on to tell us what could not be seen before, relationships among organisms that shared nothing by way of body form. Sippers of fine whiskey wouldn't be able to tell whether they were more closely related to the cereal or the yeast without using the information in genes. (Incidentally, it's the yeast). Applied more broadly, genes tell us about relationships among the majority of living things (bacteria, archaeans – superficially bacteria-like forms but belonging to an entirely separate kingdom inhabiting extreme environments – protozoans, yeast, nematodes) that have left little fossil record and in many cases are too simple in shape for anatomy or fossils to provide sufficient information.

A final key element missing from Darwin's work was the question of how selection operates on development of each individual in each generation to produce the evolution of phenotype. Darwin had a glimpse that embryos and larvae were important to understanding evolution, but he made little use of the idea. The observation at the time was that the young stages of animals resemble the adults of more primitive relatives. This is an observation first made and explained by the pioneer embryologist Karl Ernst von Baer in 1828. Von Baer discovered that animals did not pass through the adult stages of their ancestors but shared similar stages of development. The similarities grew weaker with developmental age as growing animals of different vertebrate classes diverged. Von Baer did not accept natural selection, but the evolutionary perspective on development would become a powerful idea for his and Darwin's successors.

The first major step was taken by the idiosyncratic German-born Brazilian biologist Fritz Müller, who in 1863 wrote a book whose title in the English translation is *Facts and Arguments for Darwin*. Müller noted that diverging species could develop the same way for a while and then diverge at any point in development. They could also evolve by adding successive new adult stages. This would require a compression of stages but would effectively allow a reading of the evolutionary record of an animal from its developmental stages. German zoologist Ernst Haeckel seized upon these ideas and in 1866 made them the foundations of phylogeny. His influence was enormous in research and education over the next generation. To Haeckel the study of development was important solely because it documented evolutionary history, and he thought that the fossil record was unnecessary. Although Darwin drew the first phylogenetic tree, it was a purely hypothetical scheme. Haeckel used embryology to construct elaborate trees of the relationships of real animals. One of the charming features of these trees is that they are literally drawn as trees – great knurly oaks of phylogeny. Developmental biology would not contribute much more to understanding how evolution of form took place until Haeckel's views were displaced by the mechanistic study of development.

The cosmic mystery – of flavour – lies inside the egg.

Sydney Morning Herald

THIRTEEN

Life with Sea Urchins

SEASIDE SCIENCE

My former postdoctoral advisor, Paul Gross, liked to say that "your graduate students are your friends, but your postdocs are your enemies." This was because graduate students would go off to do postdoctoral work in new areas of research but departing postdocs would want to kick-start their independent careers by continuing the research they had developed during their postdoctoral years. Thus they were virtually destined to become competitors with their own former mentors. Nonetheless, Paul was generous about allowing postdocs to take their projects with them. For several years at Indiana, I continued the study of how protein synthesis was stimulated at fertilization of sea urchin eggs. By 1978, we had published our main findings on the "masking" of mRNAs of eggs and on the regulation of protein synthesis from these mRNAs in embryos following fertilization, and I began to get restless. I was ready to start a new direction of research. I thought it was time to return to my early interest in evolution, but now with a better understanding and better methods, and a view toward thinking about evolution and development. Developmental and evolutionary biologists had diverged in their research objectives to the point where neither thought much about the role of the other. My first approach to the experimental study of the evolution of development would be based in Indiana but was boosted by priceless summers in Woods Hole, Massachusetts, where I learned about marine embryos as creatures with life histories and their own evolutionary careers. Those evolving larvae have occupied my scientific life for three decades.

By 1979 Paul had become director of the Marine Biological Laboratory at Woods Hole. This was his dream job, and he took to it passionately. He invited me to become the instructor in chief of the Woods Hole summer embryology course starting in 1980. I was thrilled to take on this venerable course, which had been offered continuously since 1893. The great founders of American embryology taught the first years of the course. These scientists, notably Charles O. Whitman, Edmund B. Wilson, Frank R. Lillie, Edwin G. Conklin, Thomas Hunt Morgan, put into play the idea of German embryologist Wilhelm Roux that the workings of the embryo should be central to the study of development – not phylogeny, which had dominated the study of embryos from the 1860s to the end of the century. Thus the MBL became an indispensable center for the new science of experimental embryology, and the severing of development from evolution. With all the excitement of probing the cellular mechanisms by which development proceeds, evolution was easily put into the attic, like a gift painting of dogs playing poker, an irrelevant and time-wasting distraction. Following Roux's lead, the Woods Hole embryologists shifted their studies to understanding how development worked as a process. Understanding the machinery of development became the goal. The impressive compound German name that Roux coined for this new science tells it all: *Entwicklungsmechanik,* which translates in English as "developmental mechanics."

The Woods Hole embryologists made the tracing of the lineages of cells in embryos their organizing principle for achieving this new mechanistic understanding, with cells as the gear wheels of the embryo. They invented methods that allowed the destinations of each cell of an embryo to be traced. The results were spectacular, revealing that the egg was organized into regions that influenced development. These different fates were mapped and then studied by experimental manipulations of cells with distinct fates to learn how they got their differences. Developmental biology had become a vigorous discipline devoted to mechanistic questions about the function of developing organisms continuing the scientific vision of the Woods Hole pioneers. By the 1890s, with the idea of development as a handmaiden for tracking evolutionary history discredited, the focus on embryos for phylogeny building was over and the evolution of developmental features was relegated to only a peripheral curiosity. This dichotomy between evolutionary biology and functional biology continues to have a disconcerting grip into our times. The present-day bias of developmental biology leans heavily on the functional side of biology, finding a place in the hearts, souls, and pocket books of funding agencies for obvious biotech and medical reasons.

The late 1880s saw the creation of seaside marine biological laboratories around the world – the MBL in 1888 was the third. The first of these indispensable windows on the sea was the Stazione Zoologica, founded in Naples in 1872; the second such laboratory was founded at Misaki, Japan, in 1887. I was fortunate to be invited to take part in a symposium in Japan held in 1987 to celebrate the centenary of the founding of the Misaki Marine Station. I don't mean "celebrate" lightly – the saki and wonderful seafood flowed freely, and Crown Prince Akihito, now the emperor of Japan, took part and mingled with the speakers at a reception. The Japanese postal service even issued a Misaki commemorative stamp. There is a beautiful link between Misaki and the MBL. On the wall of the MBL library hangs a framed note written by a famous Japanese marine embryologist, Katsuma Dan, who worked on sea urchin embryos. I first met him one summer in Woods Hole and then again in Japan when we were both part of the Misaki celebration. Although he was in his eighties, he volunteered to come to work with us in Australia, but that never proved possible.

As World War II drew to its end, Katsuma Dan was working at the Misaki Marine Station. The Japanese navy had a submarine base nearby, and he was concerned that the U.S. forces landing in Japan might destroy the marine station, thinking that it was part of the military installation. As the staff evacuated, Katsuma Dan left his note written in English for the American occupying forces: "You can destroy the weapons and the war instruments but save the civil equipments for the Japanese students. When you are through with your job here notify to the university and let us come back to our scientific home." He signed his note, "The last one to go." As a result, Misaki was spared from demolition. The commander of the American submarine squadron that occupied the bay knew about the MBL and sent Dan's note there, where it is has been displayed through the years since.

To take on the direction of the embryology course was a big honor but also an intimidating task because there was a long tradition to uphold, and a huge amount of logistics and staffing were needed to get the thing running by summer. As director I would outline the course program, inveigle the course faculty to leave their home labs and come to Woods Hole to teach, find a class teaching assistant, arrange for borrowed equipment, make sure supplies were ordered, choose the students from among the applicants, and decide on most of the invited speakers (we had one outside speaker each week). Beth and I hosted most of them. We boiled kettles of lobsters and washed dishes into the night. Hosting sometimes meant helping our visitors take sick children to the pediatrician, finding their sunglasses lost in the surf, or cooling down explosive political arguments between guests. One party actually broke up in a fiery disagreement over Reagan's firing of the air traffic controllers, and two of our guests stormed out of the cottage. Beth had left the room for a few minutes to tuck in our three-year-old son and returned to find that only one apologetic guest remained at the dinner party. Once I was awakened in the late hours by a distraught just-arrived foreign visiting speaker whose bathtub in MBL housing had no stopper. I found him one.

Amid all this, first-class science was done at Woods Hole within the context of the courses. The most famous example during my time there took place in the summer of 1982. I walked in one day and ran into Tim Hunt, an instructor in the physiology course upstairs. He was excited and said he had something to show me. It was an x-ray film with dark bands that showed radioactive-labeled proteins from a simple experiment to see if any proteins might be made and degraded during rounds of cell division at the beginning of the development of sea urchin eggs. There were. This straightforward experiment yielded the first observation of the "cyclin" cell division proteins that won Tim a Nobel Prize in 2001.

Each summer, Beth and I packed the family into the car and drove to Woods Hole accompanied by some of our graduate students driving a rental van full of lab gear. We moved into one of the summer cottages owned by the MBL, at 13 Devil's Lane. The cottage was perched on the edge of an extensive bowl-shaped depression. This was a kettle hole, a glacial structure left when a huge mass of buried ice slowly melted as the Ice Age glaciers retreated from the glacial debris pile that is Cape Cod. Now, a hundred centuries later, the quarter-mile-wide hole was filled with scrub trees and was home to an unruly gang of night-raiding raccoons. We had to tie the lids on our trash cans down with hefty nylon ropes and rubber bungee straps to keep the raccoons from prying them off with their facile hands. If their noisy efforts succeeded, I'd have to run outside into the night in my pajamas with open umbrella in hand to fend off the cheeky foes and rescue all the attractive tidbits of garbage. The raccoons looked upon my efforts with philosophical detachment.

Near the Wellfleet Bay Wildlife Sanctuary on the Cape Cod Bay side, there was a boardwalk down through the salt marsh. From there we could wade across a clear tidal stream on the way down to the barrier beach. The creek varied from knee deep at low tide to chest deep at high tide. The shallow, sloping sand beach was packed with multicolored scallop shells. The two sides of Cape Cod are only a few miles apart but lie in distinctly different water temperature zones. The summer seawater temperature at Woods Hole is several degrees warmer than that in Cape Cod Bay, because the southern part of the cape is warmed by the Gulf Stream. The Labrador Current coming from the north substantially cools the bay. The sea urchins we studied were represented on the northerly bay side by the circum-arctic green sea urchin bearing the easily spoken Latin binomial of *Strongylocentrotus droebachiensis* and at balmier Woods Hole by the warm-water red sea urchin (*Arbacia punctulata*). Farther up the Cape on the open Atlantic of the Cape Cod National Sea Shore, there were sand beaches that disappeared into the horizon.

I'm a sucker for "living fossils." The most famous of all, the horseshoe crab *Limulus polyphemus,* breeds happily on beaches near Woods Hole. Once visiting Cuttyhunk Island, the last island in the rocky Elizabeth Islands chain off the cape, we took a hike to the empty far end of the island. There is a little bay there heavily strewn with glacially rounded granite cobbles the size of pumpkins. We found lots of intact dried *Limulus* carapaces lying amid the boulders. Some were from eighteen-inch-long females. The place evoked an image of a stranding of large trilobites on some long-departed Ordovician seashore half a billion years ago, when trilobites ruled the seas. Snorkeling off the beaches at Woods Hole offered sandy bottoms and ghostly fleets of small pufferfish and ctenophores. Appealing hermit crabs occupied abandoned snail shells and scuttled over the sea floor with their awkward limestone houses. I'd meet the occasional lobster when snorkeling in rocky areas – for some reason they were always wary of associating with people.

Amanda was ten our first summer at Woods Hole. She was the ideal age for the summer programs for children and for the Woods Hole School of Science. Beth and I sometimes helped lead field trips for the kids into local salt marshes. The kids were often indifferent to lessons on mussels and salt-tolerant weeds, but they never minded wading and splashing in the wet mud. The School of Science was an enriching place for children, but unfortunately closed a few years later. Three-year-old Aaron loved going to the beach so much that the day after we had driven home the grueling eleven hundred miles from Woods Hole to Indiana, he pleaded, "Mom, lets to go to the beach." Beth told him, "We're home now. We can't go." He looked puzzled. "Why not, it's a nice beach day?" He didn't understand that going to the beach is like real estate values – it's all location.

Aside from the organizing jobs and their accompanying minor disasters, summers at the MBL were a continuous immersion in science. We had lectures and discussions in the mornings and labs all afternoon. The course offered possibilities to organize meetings to explore new developments. I took part in editing two books published in the MBL Lectures in Biology series, based on two symposia connected to the course that we organized. The first was called *Time, Space and Pattern in Embryonic Development* and was edited by Bill Jeffery and me in 1983. Bill went on to become a leading figure in chordate evo-devo, and is now the maven of blind cave fish. The second book was *Development as an Evolutionary Process,* which Beth and I edited. This book resulted from a 1985 meeting we set up at MBL around the work of a number of researchers in the evolution of development. The meeting was an early milestone in the appearance of a modern experimental science of evo-devo.

WHAT MAKES EVOLUTION GO?

I've mentioned evo-devo a couple of times, and it's time to amplify. "Evo-devo" is the affectionate name for the fusion of evolutionary and developmental biology. The aim of evo-devo is to explain the origins of the spectacular big features of evolution. This is macroevolution, or evolution at the level of major changes – new species, new body plans, the appearance of trilobites, dinosaurs, birds, flowers, butterflies, whales, and humans. Here lies the evolution of novel features, new adaptations such as eyes and limbs, even of new creatures such as horned dung beetles and humans. This is evolution on the big stage – the evolution that sells books, the evolution that infuriates creationists.

The term "evo-devo" is shorthand for something that comes from a fundamental biological observation. New body form does not evolve by adult animals of one kind shape-changing into animals of another kind. Illustrations of this adult morphing give a dramatic shorthand way to visualize change in evolution, and evolution has generally been shown this way in museums, textbooks, and especially videos, in which evolution can be set to music and motion. It makes good television, but body form arises only through the process of individual development, and so it can only evolve through modification of development in an evolving lineage from ancestor to descendant over generations. That understanding directs us to study the indirect processes whereby development and evolution are interlocked. We can investigate the process of development in living organisms and define the genes that underlie its operation. We can study how genes and development evolve over the short term, and we can investigate how genes and development evolve over geological time. This combination has the power to reveal the intimate details of evolutionary change from ancestor to descendant, and to show us how spectacular new features of organisms evolve.

It was British zoologist Walter Garstang who made clear in 1922 the vital connection between development of the individual and the evolution of a lineage of organisms. He noted that one typically envisions evolution as the adult form of an ancestor evolving into a new descendent adult form. Garstang pointed out that changes in the shape of living things does not arise this way at all. Adults do not transform into different adults. There is just the clunky generation-by-generation reproduction of new individuals. Organisms reproduce in each generation through the process of development. That involves production of an egg that develops into an embryo, the embryo to a larva, the larva to a juvenile, and the juvenile to the reproductive adult. In development, form arises as a process directed by the genes of an organism operating within the cellular machinery of the egg. The pathway of development is thus mapped out within the range of environments, such as temperature, that allow normal operation of the cells and genes. In some cases, different environments draw forth alternate forms of development. Evolution also means change, but it works differently. The changes we see over time in evolution are not part of a predictable and reproducible agenda but are the result of change in developmental processes each generation. Evolution is contingent and it depends on undirected changes in the DNA code. The results of selection that favors change are the increase of variants that have different genes and developmental machinery from the ancestor.

The link between the two distinct processes of evolution and development arises because form takes place in each generation through development from egg to adult, and that process of development must itself evolve over geological time such that forms intermediate between ancestor and descendant are produced. The picture of the role of development envisioned by de Beer was that new kinds of animals could arise by relative changes in the timing of events in development. For example, suppose that a descendant species grows its legs sooner than its ancestor. It may come to have longer legs as a result of simply shifting "developmental gears." No big change in mechanisms is required, just how those mechanisms are deployed in time. This kind of evolutionary change is called "heterochrony," which means a shift in the timing of developmental events relative to one another. The term itself dates back to the late nineteenth century and was coined by Ernst Haeckel, who theorized that all evolution of animals was driven by the gain of new developmental stages. Each new adult stage would be added at the end of development, with the previous stages retained in abbreviated form. This idea, called recapitulation, was considered by de Beer to be too narrow, and he expanded on Haeckel's idea that new stages could be inserted earlier in developmental timing as well. De Beer proposed that timing changes in early development would have more profound evolutionary outcomes and defined several kinds of displacement in time of one process in development relative to another in timing.

Heterochrony was so compelling an idea that it dominated studies of developmental evolution throughout the early twentieth century. At that time, the kinds of macroevolutionary changes that engaged biologists included the attempt to explain the evolution of most everything, including the human body, by means of heterochrony. So we were seen essentially as adults sexually, but equivalent in body form and learning flexibility to young apes. Other studies focused heavily on heterochrony in larval evolution. This was the basis of Gavin de Beer's book, *Embryos and Ancestors*. Although interesting organisms and evolutionary-developmental phenomena were studied by investigators of the period, the lack of sufficient new concepts and tools meant that the enterprise petered out, and by the 1970s it was in eclipse.

There are actually two forms of evolutionary biology. The first is called "microevolution," and it is the kind of evolution most studied. Microevolution focuses on the processes that operate within populations, such things as how mutations arise, how variants of genes spread in populations under selection, and how selection functions in a variety of circumstances. Evolutionary work with a microevolutionary outlook has been enormously powerful in unraveling what the basic machinery of evolution is on the scale of within a species and at each tick of the evolutionary clock, both in the lab and in nature. This approach has been hugely important and influential. Its success has unfortunately led to the widespread use of an overly narrow definition of evolution (often quoted in text books): "Evolution in a population is change in gene frequencies from one generation to the next." This is not an incorrect definition, but it surely is grossly incomplete, and it makes the subject of evolution sound like a deadly yawner to most people who have in mind the grand sweep of evolution across geological eras.

Microevolution held sway for most of the twentieth century, until new techniques of developmental genetics reached sufficient sophistication to penetrate to the heart of development and let us delve into the evolution of organismal form over long geological times. This kind of big evolution, which is coming into its own as an exciting and dynamic field, is macroevolution. So how does it relate to the other evolution, the microevolution among populations within species studied by most evolutionary biologists? The assumed relationship is that microevolutionary mechanisms operating over long periods of time produce macroevolutionary changes. Basically, the concept is that if you can walk ten feet across your lawn, given enough time you can walk from Maine to California. However, others have wondered if perhaps macroevolution might sometimes be more like going from California to Australia, an activity different from walking. If that were so, some major changes might involve mechanisms that are qualitatively different from those revealed by microevolution. Strangely, this dichotomy of ideas of how novel features arise in evolution is still not resolved, and it might be a false dichotomy.

A celebrated study, the decades-long research of biologists Peter and Rosemary Grant on the famous "Darwin's" finches of the Galapagos Islands, has provided the most compelling picture of how microevolution might lead to macroevolution. The body of work accomplished by the Grants and their students is a saga of intensive work, monitoring for years each individual in the entire population of a finch species on a single island. The Grants and their students spent six months of every year on a small desert crust of an island called Daphne Major. There they banded, released, and followed the fates of every finch on the island through thick and thin. In the 1970s, an extraordinary period of drought and rain allowed them to observe that in tough times, when small seeds disappeared, only birds with large beaks survived, and they passed on the trait of large beaks to their progeny. Selection operated in the opposite direction in wet years and was documented again; in wet years, only small beaks were selected for and passed on. Their most spectacular result was that selection for beak size can be strong, and that evolutionary responses are correspondingly pronounced and rapid. The Grants have shown that all components of Darwinian evolution are present. There is variability. It affects survival. It is heritable. It is selectable.

The problem for extrapolation of the studies on Galapagos finches to the macroevolutionary level is that nothing novel was generated. Variations in beak form and function were the result of variations on a developmental theme already long established. Given that, evolution of new forms of life might be posited to involve more than the microevolutionary processes usually studied, because no change in shape is required for microevolution to have occurred, whereas it is the large-scale change of shape over geological time that most people recognize as evolution – macroevolution. The transformations in form seen in macroevolution need a link between selection and the processes that produce form. Work by Arhat Abzhanov and a team of developmental geneticists collaborating with the Grants revealed regulatory genes that when expressed at high levels cause thicker beaks to develop and that may lie at the business end of selection for large beaks in Darwin's finches. Natural selection operates in nature, and we are now beginning to link it, not to some mysterious "heredity" but to changes in the action of particular genes. The objection that microevolution is not well linked to development is disappearing, but this does not prove that macroevolutionary changes might not involve other processes.

Once a role for the process of development in evolution was recognized, a new way of studying some of the great problems in macroevolution was illuminated. Animals could be seen as very non-machine-like entities in which the complex properties of the final adult form arise from growth and transformation within a system such that morphological complexity increases from the far simpler eggs and embryo stages that precede the adult. Evolving organisms have an even odder feature: the evolutionary modification of the living developing system while it is running. Development in an evolving lineage sees processes remodeled, parts of the developmental sequence lost, changed in time and place, and even new processes gained. The evolution of development and the genes involved in developmental processes assume major explanatory roles. Evolutionary conundrums such as the origin of multicellularity, the origin of eyes, appendages, and segments can be understood as lying in evolutionary modifications of development. Our ability to penetrate these problems has grown along with the power of the experimental analyses of development, notably from developmental genetics in the past two decades.

A revival in thought about evolution and development began in the 1970s as developmental biology itself was undergoing its renaissance. Stephen J. Gould's book *Ontogeny and Phylogeny,* published in 1977, was stimulating to a renewal of interest in evolution and development. Curiously, although scholarly, readable, timely, and inspiring, the book was actually backward looking in theoretical concepts. Gould's intent in much of the book was historical tracing not of all of evo-devo but of the one historically dominating idea, heterochrony. Gould refined the basic types of heterochronies. Although Gould criticized outdated ideas about heterochrony, he retained the broader concept as representing the major pattern and means of developmental evolution. He attempted to provide a sort of analogy to represent the kinds of heterochronies through a set of developmental clock diagrams. Gould's diagrams could only remain as metaphors until new tools and concepts began to be applied. Heterochronies occur, but their underlying mechanisms are largely unknown. Furthermore, we measure all of development along a time line, which means that inevitably we will observe many evolutionary changes as heterochronies even if the primary mechanisms have nothing to do with control of timing. This is not to say that heterochrony is not important, but that we need to integrate it into modern ideas of how development is regulated.

Evo-devo is a strictly modern term for the descendant of an enterprise that started with the nineteenth-century research notion that the problem was phylogeny and the approach was embryology. From the 1920s to the 1950s, considerations of development in evolution reflected the growth of sophistication in biology, such that the problems focused on heterochrony, evolutionary morphology, and homoeosis (evolutionary changes in identity of segmental structures, e.g., transformation of one kind of appendage into another). The approaches were morphological developmental biology, embryology, and genetics. By the late twentieth century asking more penetrating questions about evolutionary processes in development, the role of gene regulation, the origins of novel features, and the origins of animal body plans became possible.

Embryos undergo development; ancestors have undergone evolution.

Sir Gavin de Beer

FOURTEEN

Embryos Evolving

HOW TO DO EVO-DEVO

At Indiana I began thinking about evo-devo as something that should move beyond its mid-twentieth-century form. Over a period of time I developed a parts list of essential elements we would need in order to be able to both ask and answer questions about the evolution of body form. It seemed that it would take a synthetic combination of ideas from several disciplines. These eventually would include developmental genetics, molecular evolution, and genomics on the mechanistic side. But more was needed. The historical disciplines of paleontology and phylogeny would allow us to include the events of the long ago evolutionary past in our analyses. Paleontology would allow us to envision ancestral conditions and patterns of change, and evidence from phylogeny would allow us to map a history of descent. The second essential element was a suitable research organism for addressing experimental questions. That would require an organism that presented concrete evo-devo problems stemming from its evolutionary history and was amenable to experiments designed to answer those questions. The third was to seek mechanisms of evo-devo that were comparable in explanatory power to those of modern developmental and molecular biology. This would allow us to understand the mechanistic bases for particular evolutionary changes. I considered phylogeny to be crucial to the experimental effort, because it is only by incorporating the pattern of descent of evolutionary changes that evolutionary changes in development can be interpreted as arising once in an evolving lineage or evolving independently in separate lineages. We should be able to set mechanistic explanations into these phylogenies.

In cases of convergent evolution of new features, phylogeny would let us unravel the choices nature has in independently responding to similar selective pressures. That would take new phylogenetic data to accomplish. So my fourth objective was to use molecular biology to infer the phylogenies of the major animal phyla to get a better framework of relationships than previous data had allowed. These goals did not spring fully formed to mind, and I was naïve about much of what had to be done. In 1975, I tried to rigorously organize my thinking by starting a course in evo-devo for graduate students (the term we first used was devo-evo, but by consensus that eventually flipped to evo-devo because "evolutionary developmental biology" was less clumsy than "developmental evolution"). I also began writing a book to explore and synthesize the components of a discipline, and I started what was at first an uncertain quest for an experimental system that I could develop to ask evo-devo questions.

As far as I know the start of my graduate course seems to have tied an analogous course at Berkeley as one of the first evo-devo courses taught anywhere. I asked my geneticist colleague Thomas Kaufman to join me in teaching that class as part of writing a book. Thom was beginning a massive study of the homoeotic genes that control the layout of body development, and he would assume a major role in a revolution of our understanding of the genes that regulate development in the animal kingdom. We taught the 1977 class together and drafted an outline for what would become the book *Embryos, Genes, and Evolution,* published in 1983. I hoped that we could provide a guide to the making of a modern evo-devo discipline. The publisher, Macmillan, gave us a small advance and some money to pay a typist for transcription. We opened a joint account at a bank downtown so either of us could write the typist's checks as needed. We had both our names put on the checks, which amusingly raised a few eyebrows at the bank when we went down together to open a joint checking account. Beth illustrated the book with a large series of superb line drawings.

Embryos, Genes, and Evolution was important because it was the first attempt to link evolution with the new developmental genetics, and it was the first book to map the course that evo-devo would take. We integrated the older traditions of the study of evolution of development with new experimental approaches being created in developmental biology. At the time there was no actual discipline of what came to be called evo-devo. There was an earlier tradition of embryologists studying evolution of development, but because of technical limitations, that field had eventually petered out. We had to borrow most experimental examples in the book from work done for other reasons than evo-devo. We hoped to stimulate development of a new field that would create its own research questions and experimental systems. We were right in our pointing out of future directions, although our examples seem primitive now.

A DAY IN EAST BERLIN

In 1980, as we were writing the book and I was thinking about redirecting my own laboratory research, I was invited by John Bonner to take part in the 1981 Dahlem Conference he was organizing to be held in the Dahlem district of West Berlin. This was to be a broad multidisciplinary brainstorming session about the nature of developmental evolution. One of the notable features of the meeting was the setting. Berlin is now a vibrant, unified city and the Cold War is only a faint memory. That was not the case when we met there in 1981. I landed at the old Tempelhof Airport, which would be closed in 2008. The plane banked low over the Wall that separated the two Berlins. This medieval and claustrophobic prison for an entire city stood in contrast to the meeting site, the spectacular glassed top floor of a new tall building in West Berlin. The street below was the Kurfürstendamm, a glitzy shopping avenue of expensive cars and prostitutes showing off their short skirts and high white boots. The streets at night could be unexpected in other ways too. I was awakened in my hotel a couple of times by the sounds of rumbling engines and heavy metal screeching on heavy metal – a procession of American tanks on night maneuvers through the city streets, playing mind games with the Russians and East Germans. East Berlin was visible from the conference site windows.

One of the attendees was Gunther Stent, the molecular biologist who in 1968 had notoriously (and prematurely) declared that molecular biology was dead. He thought that only the study of consciousness could offer any hope of finding new laws of biology. I wondered why he had come, and I suspected he was looking to see if novel ideas lay in this attempt to blend separate disciplines. Stent was a native Berliner who, as a Jewish teenager, had fled Germany. He knew the city well and was a good guide and raconteur. I took a stroll with him around the old embassy quarter, dead and darkened empty buildings near the wall, which his vivid stories returned briefly to the long-ago life as recalled from his youth.

Later in the meeting, a number of us elected to take a bus trip into East Berlin. We had to turn in our passports a few days in advance of leaving for the edification of the East German authorities. Delegates from Poland were not allowed to come along, because Poland was already regarded as ideologically unreliable, but Western delegates were welcome. On the appointed day we boarded the bus and headed for Check Point Charlie, the crossing point. There the East German border guards made us get off the bus and stand in a gym-class-like file so they could compare each of us to our passports. They searched the bus for contraband like newspapers or magazines. Then we picked up our official tour guide on the east side of the checkpoint, and off we went. East Berlin was shockingly drab, with buildings that had been bombed out in wartime still standing here and there. We went first on a mandatory visit to the Soviet War Memorial with its massive grave mound and oversized, clunky statues of burly Russian soldiers and grieving mothers. One sarcastic colleague aboard the bus noted that our German guide did not seem especially distressed at the thought of dead Russians.

We requested a stop at the Humboldt Natural History Museum, which is the home of the Berlin *Archaeopteryx* and the largest dinosaur skeleton ever mounted, the East African sauropod, *Brachiosaurus.* The museum was one of the showpieces of imperial Germany's scientific preeminence. It by chance had survived the saturated bombings of the city. One of Carnegie's casts of the great Pittsburgh dinosaur *Diplodocus* was mounted next to the *Brachiosaurus* for scale. The *Diplodocus* looked small by comparison – a point of pride for German paleontology. The museum in 1981 exuded an air of paranoia and inaction. The governing of tours was rigid. In trade for our requested visit to the natural history museum, we lost the chance to stop in at the Pergamon Museum's vast collection of ancient art, because it was not permitted to miss propaganda sites such as the war memorial. We returned to Check Point Charlie and again had to disembark for another going over. The underside of the bus was searched using mirrors on wheels inserted under the vehicle to make sure no one was trying to flee the workers' paradise by clinging to the undercarriage. There was no clue that the wall and the whole charade of East Germany would come down in less than a decade. In 2010 a retrospective of the 1981 meeting was held at the Max Planck Institute for the History of Science. Beth and I took part. We would finally see the new city and both the revivified Natural History and Pergamon Museums.

The conversation about evo-devo drew scientists from several disciplines, but in 1981 evo-devo had not then yet jelled as a discipline. Discussions of mathematical models of the rules of leg development, of genes and development, of life history, and of macroevolution were exciting and mixed well in our sunny glassed-in tower top. The personalities who would take part in the coming field of evo-devo were emerging, but this meeting was still just the dipping of toes into the water. Two classic concepts would dominate our ideas and discussions about how development and evolution would mesh. These were heterochrony and the notion of "developmental constraint," the latter term referring to limits on what selection can do in the face of the genetic and developmental "rules" of an organism. Perhaps strangely, the questions raised by these concepts have even now not been answered. In fact, the classic "mechanisms" of heterochrony and constraint would be left behind during the next few years, as new experimental approaches based on developmental genetics would come to the fore. Much of the later progress also would come from fields not as well represented in the 1981 meeting, notably phylogenetics and paleontology.

Participants in the 1981 Dahlem Conference probably had as their most important question, that of the potential novel roles for development in evolutionary theory. That is, does development provide any new mechanisms for evolution that were not included in a view of evolution based on natural selection acting on small-scale genetic variation? Darwin realized that to propose a real theory of "transformation," he would have to provide a general mechanism. Natural selection offered that core mechanism of evolutionary change, and it has endured over a century of testing. The Dahlem Conference probed the idea that the rules of developmental biology might contain a novel mechanism, at a distinct level from selection, for generating evolution. The forms that contributions by development might take broke down roughly into three possibilities.

At one end of the spectrum in the conversation was the conventional view that developmental processes would evolve according to the classic rules of microevolution. Small mutations would occur at random in genes regulating development. The phenotypes generated by them would be subject to selection. Evo-devo would offer much information about how development had evolved but would not produce any novel evolutionary mechanisms. In 2000, Armand Leroi would cogently argue this position. He incorporated a discussion of new ideas on the magnitude of effects of mutations. Modern work indicates that mutations are not all of small effect. There is a scaling of mutational effects. Leroi argued that accommodation to a broader range of mutational effects would allow for microevolution to supply the change seen in the long scale of macroevolution. This view might be called the "no new mechanisms" position. In this scenario, evo-devo would enrich our knowledge of macroevolutionary patterns, but not add so much to processes, which would still be those of microevolution.

Steve Gould argued a controversial viewpoint at the opposite end of the spectrum. He supported the idea that the rules of development might produce distinct stable states not easily disrupted by developmental mutations. Gould used as an analog a solid polygon. Suppose it has eight sides and sits on one of them. It is stable to perturbation up to the point where enough force is applied to flip the polygon over. An infinite number of new states are not possible. Any one of the remaining seven sides offers a new stable position – but that's all. If developmental evolution were limited in an analogous way, selection would only have a few constrained new solutions available to it. These new solutions would be developmentally stable and move the developmental system from one stable built in "peak" to another. This kind of idea was not new. A generation earlier, Richard Goldschmidt had taken a view that argued that changes from one well-adapted form to another by microevolutionary changes would disrupt adaptation and thus be a weak mechanism of morphological evolution. He suggested from studies of the effects of mutations in "homoeotic genes" that large-scale evolution could be driven by chromosomal "rearrangements" of such genes that allowed jumps in body shape to take place. He referred to these genetic jumps as "hopeful monsters."

Per Alberch, a gifted young scientist who like Gould would be lost to us by death before the 2010 replay of the Dahlem meeting, promoted a heuristic concept of constraint by a simple graph of the traits of individual species as distinct clusters of points on an imaginary landscape. The variability of different species would fall in discrete clusters of points. The coherency of clusters could be maintained either by selection or by internal rules of development. Suppose the graph was of length versus width. Turtles are pan-shaped, and snakes are, well, snaky. Turtles are nearly as wide as long – snakes never even close, even when well fed. Species of turtles and snakes would cluster into different distributions on the graph. Alberch suggested a thought experiment that separated the effects of selection from internal "constraints" in maintaining discrete shapes. If there were no constraints, and selection were removed, over time the points would diffuse, because there would be no internal force to maintain particular developmental forms. Conversely, if constraints existed, the points might move but would remain in discrete clusters. That is, developmental regulatory systems might come to constitute organization of a kind that would allow some kinds of evolutionary modifications but limit others. Unfortunately, this is a fairy godmother experiment. No test like this can be done in the real world, although Alberch devised experiments that showed that developmental constraints apparently lie in the evolutionary loss of fingers and toes.

At the conference some developmental biologists of a theoretical bent supported the notion of underlying rules for developmental patterning of such things as the digits of the hand. They suggested that some digit configurations would not be developmentally possible and thus not found in nature. The weakness of the rules argument is that it's hard to test the reason for something's absence. It might just be due to chance that some path was not followed, or to conventional natural selection.

These particular arguments for constraint have been weakened further by the discovery of how patterning genes actually act to produce a developmental system.

Consistent with the ideas argued by Gould and by Alberch at Dahlem, Gould and Niles Eldredge had in 1972 interpreted the geologically sudden appearance of new species in the fossil record to represent what they called "punctuated equilibrium." This model proposed that new species arose rapidly as a result of mutations that flipped an isolated population from one stable state to another. A fast and genetically simple change could result in a new species because alternate stable states existed. For example, the numbers of rows of lenses in trilobite eyes was seen to change suddenly in the fossil record. Again, questions arise. What if new and old numbers of rows of eye lens were equivalent in function? An evolutionary change of this type could lie outside of selection and instead be governed by developmental rules.

It is unlikely that any developmental feature that arose through selection should become so tightly constrained that it could not be at some point deconstructed by selection. Is there a "third way" to think about developmental-genetic biases in evolution that doesn't live at one end or another of the spectrum between microevolution by selection and rigid developmental constraint? Can constraints truly influence evolution? I think that they do, but not in nearly so dramatic a way as suggested by the structuralism of Gould's polygon. There really is no reason to doubt that selection is at work in the evolution of development. The notion of constraints becomes one of several factors that may bias the response to selection. These would include the evolutionary history of the system, including the patterning that precedes the constrained point in development. The possibilities also include the degree to which developmental modules became linked to other major processes or remained independent. Developmental constraints will not force rigid limitations on selection so much as biases in the production of selectable variation, but in no organism can variation be unbounded because only a limited set of genetic and developmental information exists.

Constraint further means what is reachable depends on what genes and gene regulation systems are present. That limits what mutations can occur. Not all conceivable ones will be possible – no angel wings

A gallery of sea urchins we have used in evo-devo research: (*top left*) the animal that launched the years of our research in Sydney, *Heliocidaris erythrogramma;* (*top right*) the closely related species, *Heliocidaris tuberculata;* (*middle left*) *Pseudoboletia maculata,* another Sydney sea urchin used for studies of cross species hybrids; (*middle right*) its close relative, *Pseudoboletia indiana;* (*lower left*) the sea biscuit urchin *Clypeaster rosaceus* from Panama, which gave us a model evolutionary intermediate in the evolution of larval development; (*lower right*) an uncomfortable denizen to have in our research tank, the toxic-to-the-touch sea urchin *Toxopneustes pileolus.* The petals of the little "flowers" on the surface are tipped with sharp venomous syringes. *Photographs, Beth Raff.*

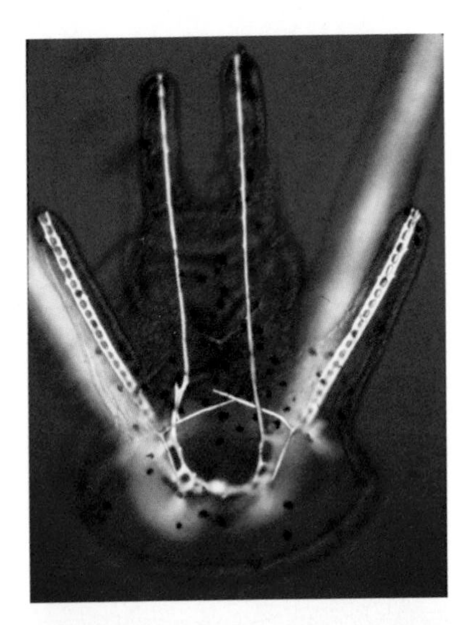

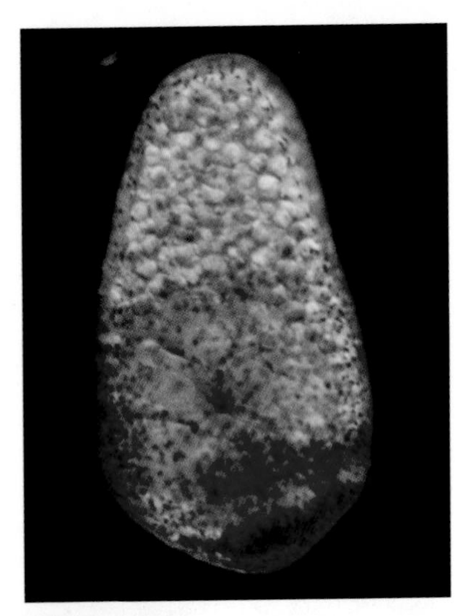

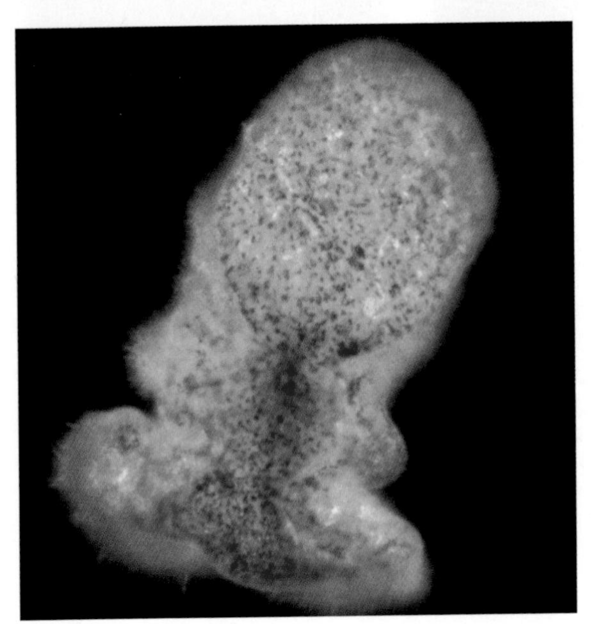

The beauty of sea urchin larvae, and their evolution. (*top left*) The pluteus larva of *Heliocidaris tuberculata*. The long arms have calcite skeletal rods. The arms are involved in swimming and food capture. (*top right*) The fast developing, non-feeding larva of *Heliocidaris erythrogramma*. (*bottom*) A cross species hybrid larva between an *H. erythrogramma* egg and an *H. tuberculata* sperm. *Photographs: H. tuberculata* larva, *Ellen Popodi. Other photographs, Raff Lab and Beth Raff.*

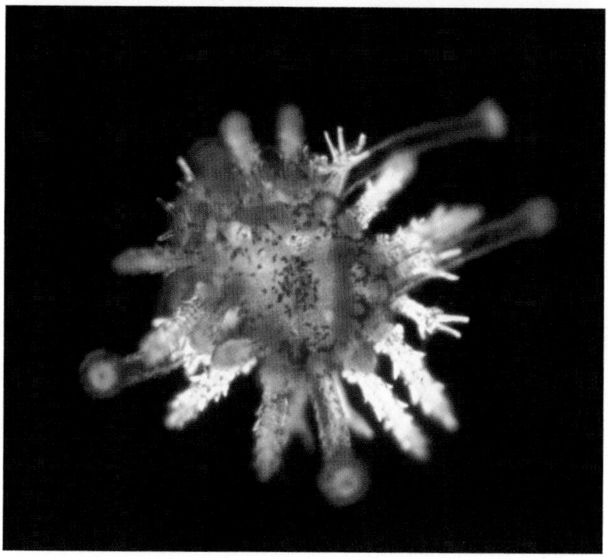

(*top*) An adult *H. tuberculata* gazes out to the lab from our research aquarium, Sydney. (*bottom*) A newly metamorphosed hybrid sea urchin. *Photographs: Beth Raff.*

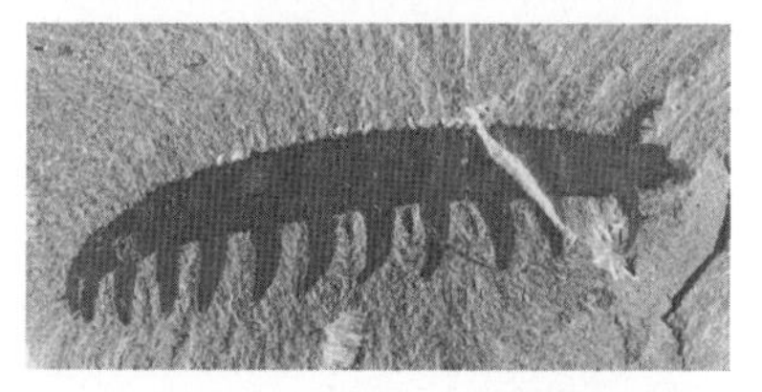

Hunting the living fossil, peripatus, in the Blue Mountains of Australia: (*top left*) live peripatus in a decaying log: *Euperipatoides rowelli,* Tallaganda State Forest, near Canberra. *Photograph, Robyn Stutchbury.* (*top right*) Noel Tait and Beth Raff dissect a decaying log. *Photograph, Rudy Raff.* (*middle right*) A fossil Cambrian "peripatus," *Aysheaia pedunculata,* from the Burgess Shale of Canada. *Courtesy of Thomas Jorstad, used with permission of the Natural History Museum, Smithsonian Institution, Washington, DC.* (*bottom*) Noel Tait's bush lunch. *Left to right*: Greg Rouse, Noel Tait, Lennart Olsson, Rudy Raff. *Photograph, Beth Raff.*

View of Brachina Gorge and the Flinders Range, South Australia, looking South. The jagged teeth just left of center of the image are the rocks that bear the famous Late Precambrian Ediacaran animal fossils. *Photograph, Rudy Raff.*

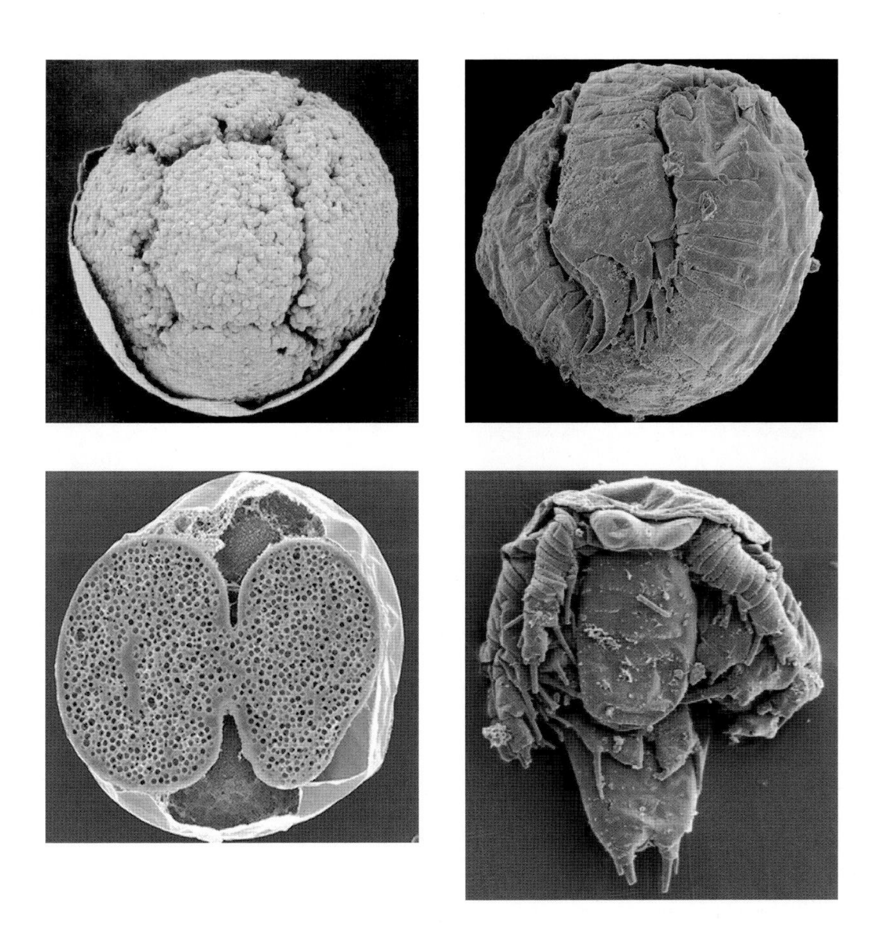

(ABOVE) Soft-bodied fossils and modern experimental analogs used for the study of the role of microbes in fossilization. (*top left*) A Late Precambrian fossil embryo. *Image courtesy of Shuhi Xiao, reprinted by permission from Macmillan Publishers Ltd: Xiao et al.,* Nature *391:553–558 (1998).* (*top right*) A fossil Cambrian "worm" larva, *Markuelia. Image courtesy of Philip Donoghue. Used by permission of John Wiley and Sons: Donoghue et al.,* Evo. Devo. *8:232–238 (2006).* (*bottom left*) A four-cell *H. erythrogramma* embryo split to reveal cell interiors by scanning electron microscopy. (*bottom right*) *Bradocaris,* a fossil Cambrian crustacean larva, preserved as apatite. *Image courtesy of Dieter Waloszek.*

(FACING) Visiting the dawn of the animal world. (*top left*) Resting on the rippled Ediacaran sea floor in Brachina Gorge. (*top right*) Ediacaran fossils were formed when storms dumped sand over the living animals. Fossils are preserved on the bottom of rock layers. Here is a *Dickinsonia.* The peculiar elephant skin texture is the remains of an algal mat that covered the sea floor. Specimen in the South Australian Museum, Adelaide. (*lower left*) Sponge-like archeaocyathids, one of the oldest Cambrian animal fossils of Brachina Gorge, about 20 million years younger than the Ediacara fossils. (*lower right*) An Ediacaran frond animal, *Charniodiscus.* Specimen in the South Australian Museum, Adelaide. *Photographs, Beth Raff.*

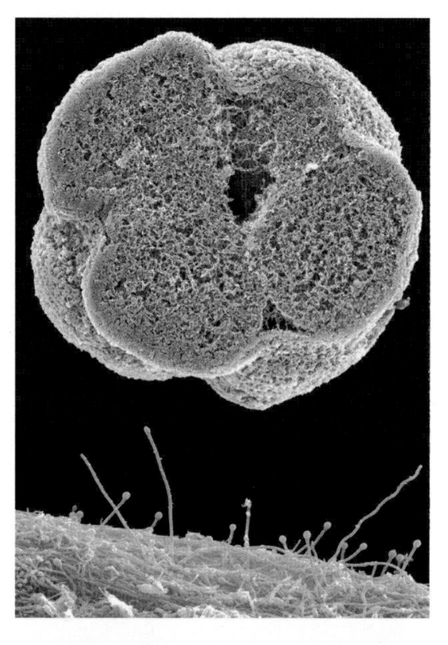

An embryo exposed to seawater bacteria for five days. All parts have been replaced by bacteria that consumed them, but at the same time replaced and preserved cell structures as a pseudomorph. This now has the potential to be replaced by the mineral apatite on the way to fossilization.

The surface of an embryo on which a rich biofilm topped by odd bacterial shapes is growing.

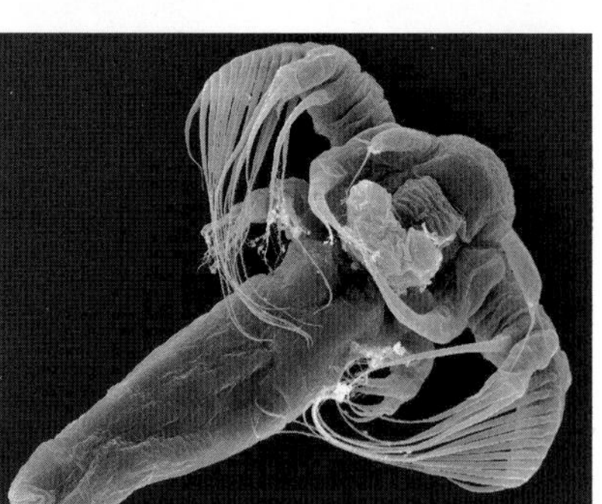

A brine shrimp.

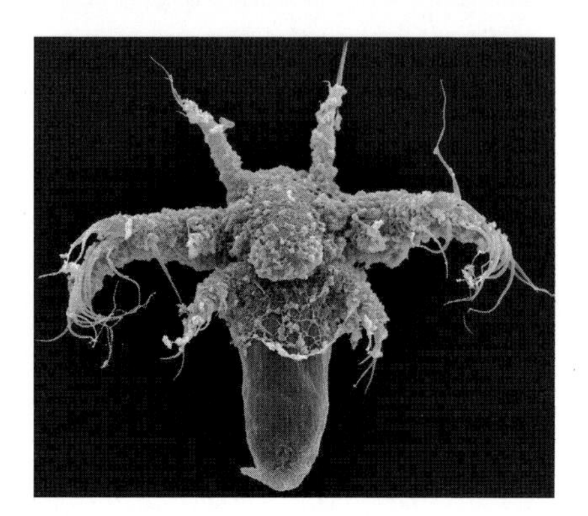

A brine shrimp that has been allowed to undergo microbial action. The shabby sweater–looking covering is a bacterial biofilm.

Scanning electron micrographs of modern embryos and Artemia, *F. R. Turner, Raff Lab.*

gracing vertebrates. In *The Shape of Life,* published in 1996, I suggested that modules are the units in which developmental genetics undergoes evolution. Modules are semi-autonomous domains of a developing animal, like limb buds or developing organs, or a gene network, each with a discrete pattern of gene expression. Modules become important to constraint, but not because they can't evolve. They do. However, in the course of evolution of complex animals, new modules get added, and they become integrated into systems of cell-cell communication, adding structural and genetic complexity at a higher probability than the invention of something entirely new. Mutations in genes involved in highly interactive systems may have a lower probability of evolving than those with fewer connections. Thus, elements that pattern deep body plan features are stable because they underlie entrenched integrative roles. A system may become so well integrated that the probability of a deep reorganization becomes less likely but not impossible. Animal phyla are examples. Phylum body plans, for example, vertebrate, arthropod, mollusk, or echinoderm, were established early in the Cambrian. They have evolved since, sometimes strikingly, but the basic ground plans of each phylum have remained for 500 million years. Evolution of form once a body plan is established may be highly important to selection in the world of competition but superficial in the level in which features of the organism evolve. Vertebrate legs are all similar in basic limb bud development, but the ultimate results vary enormously from lizard to antelope to bat.

REINTRODUCING PHYLOGENY INTO EVO-DEVO

In the early 1980s, when Kaufman and I were working on *Embryos, Genes, and Evolution,* the field had not developed sufficiently to understand how significant a deficiency it was that we were using the decades old phylogenetic trees of animal evolutionary descent. These trees had been built around embryology and comparative anatomy using older methods of inferring phylogeny. The ideas of animal relationships framed in those venerable trees in the 1940s and 1950s had become frozen into dogma in prominent invertebrate zoology textbooks, but they appeared increasingly dubious. The importance of an understanding of phylogeny came into focus as evolutionary biologists realized that it was impossible to discuss evolution of body features without tracing their history in organismal lineages. About the time our book was published in 1983, two revolutions were taking place in phylogeny, cladistics and molecular phylogeny, which makes use of gene sequence data.

Cladistics, a new way of assessing relationships, was devised by German entomologist Willi Hennig in the 1950s. This is a method of using what are called "shared-derived features" of organisms to understand the splitting patterns of evolution. The idea is to unite organisms on the basis of features shared by evolving lineages. Thus all vertebrates share a vertebral column, which unites them. In the resulting tree of vertebrate evolution, the vertebrates branch from such close relatives as starfish, which are more basal. That is, they branch lower in the tree. Higher up, the family tree of vertebrates is composed of nested sets of lineages that have features in common to the exclusion of lineages more basal to them. So sharing jaws unites fish, amphibians, reptiles, birds, and mammals but excludes the most primitive vertebrates, the jawless lampreys and hagfish, which branch below all the vertebrates with jaws. So it goes with each branching. Sharing a derived kind of egg called the amniotic egg unites reptiles, birds and mammals, which branch higher up in the tree from their closest relatives, the amphibians. Cladistics provided a novel and powerful way of inferring evolutionary history and revitalized the study of phylogeny. All of evo-devo relies on phylogenetic data. We cannot understand the evolution of a structure if we don't know what its evolutionary history is. From what precursor did it arise? When? How fast? Did it evolve more than once? Did it require similar or different genes in its evolution to a convergent feature? As a result, evo-devo papers now abound in opulent groves of phylogenetic trees.

My interests in molecular phylogeny grew from wanting to find answers to two questions that had always seemed impossible to answer: How did animals arise? And how did they evolve into the distinct body plans of the phyla? I looked at two kinds of approach to this as an evo-devo problem. First, we would use molecular phylogeny to gain a sketch of what relationships among animal phyla looked like. If cladistics provided one revolutionary change, the growing use of gene sequencing as a new source of information for estimating the relationship of organisms to each other provided another. Gene sequence data have the remarkable property of being independent of the shape of organisms. Using anatomy we can infer relationships only among organisms that have enough similarities in body plan to give shared derived morphological features. With DNA sequencing we can go beyond anatomical similarity to evolutionary comparisons where essentially no morphology is shared. We can infer relationships of animal phyla to one another where similarities of form peter out, such as asking if humans are more related to octopus, insects, or starfish. We could even ask about relationships between animals and plants, and even between animals and bacteria. This is a mind-expanding concept – and one that could be tied to methods being developed for inferring molecular phylogenetic histories. Data from genes, a source not dreamed of in Darwin's time, would confirm the common descent of all organisms on Earth.

As I began to work on molecular phylogeny, I realized that using new data and methods could give new insights by allowing us to test venerable evolutionary relationships that had become fixed in the minds of zoologists. Gene-based phylogeny also allowed us to revise our understanding of another aspect of evolution, morphological convergence. Convergence is what we see when unrelated evolutionary lineages independently evolve similar features. But convergence isn't just an absurd mistake in how we read the record. Convergence tells us how distinct organisms can evolve similar solutions in response to similar selection pressures. Genes offered a way of separating the determination of relationship away from anatomical features. This approach would allow the laying bare of homology, the evolution of a feature from a common ancestor, versus convergence of independently evolved features. Knowing the difference is crucial to unraveling evolution of body features and the history of diversity of form we see in nature.

The genes we used for the study of evolutionary events that took place deep in the past were those encoding ribosomal RNAs, which are a structural element of the ribosome, the fundamental machinery that translates mRNAs into proteins. Ribosomal RNAs are universally present in living things on Earth, and because their function is both essential and conserved, they evolve slowly. Carl Woese, a genius of molecular phylogeny working at the University of Illinois, pioneered their use in phylogeny. He stuck for years with a heroic effort that few thought would succeed and was able to start disentangling the evolutionary history of bacteria. His most spectacular discovery would be an entirely unsuspected kingdom of life, the bacteria-like in shape but biochemically weird Archaea.

In 1985 Beth and I teamed up with our evolutionary microbiologist colleague Norman Pace and other colleagues, notably, Kate Field and Gary Olsen. Norm was developing methods for sequencing the ribosomal RNA of bacteria that revolutionized the ease and speed of studying the phylogenetic relationships among bacteria. These methods would give us the means of solving relationships among the animal phyla independent of form. Existing concepts of animal phylum relationships were based on embryology and anatomy dating to the 1950s. They had not been tested before by use of molecular data. We published our results in *Science,* in 1988, and although it would take a decade or more to fully frame the new phylogeny, our work showed that the old ideas of the most basic level relationships needed revision or dumping. For example, we used phylogenetic information found independently of considering body form to test the long-accepted hypothesis that annelids (segmented worms) were the evolutionary sisters or even ancestors of the arthropods. Arthropods such as insects, shrimp, and millipedes share segmented bodies, each, at least primitively, bearing an appendage. Annelids (earthworms and all those segmented marine worms) also have segmented bodies with appendages. So did segmentation evolve once or twice? Classically in zoology, arthropods and annelids were put into a single group called "Articulata." Our gene sequence results in 1988 exploded that concept, ending a confident, century-old hypothesis that annelids gave rise to arthropods. That historic concept ultimately proved to be a house built on lime jello.

Later work I did with UCLA biologist Jim Lake and other collaborators in 1997 allowed us to recognize that arthropods and their relatives formed a novel clade that Jim christened Ecdysozoa, because all the member phyla share the property that they shed their cuticles in order to grow, a process called ecdysis. Surprisingly, the relatives included the nematodes, round worms with little in body shape to connect them to arthropods. The annelids were sliced from their classically imagined sister, the arthropods, and had to join their newly realized close relatives, the mollusks, in yet another grand reorganization of animal relationships. The subsequent molecular phylogeny studies of many labs has provided more precise hypotheses of the history of animal evolution and in essence revealed time's arrow. The revised relationships would be crucial to understanding evo-devo.

GENES AND THE SHAPE OF LIFE

Strangely, none of us interested in evo-devo in the 1970s entertained the thought that regulatory genes might be similar across phyla and thus across body plans. Recall that arthropods, worms, and jellyfish are not built like us. The great twentieth-century evolutionary biologist Ernst Mayr in his 1963 book *Animal Species and Evolution* observed that it would be futile to think that genetic homologies existed between the phyla. Mayr had logic, not data, to support his thinking, but his views were highly influential. The pioneer evolutionary molecular geneticist Allan Wilson in the 1970s investigated differences in evolution of so-called structural genes versus regulatory genes and from his data concluded that regulatory genes evolved rapidly. I think all these factors, including Wilson's reputation and highly visible results, checked anyone from considering the possibility that regulatory genes could be shared as distantly as by animal phyla. In addition, the idea of developmental regulatory genes was extremely vague in that no one had characterized any of them. Cal Tech geneticist Ed Lewis had glimpsed the solution in the 1970s. Lewis would later win a Nobel Prize for his work in *Drosophila* developmental genetics. He studied homoeotic genes that controlled the identities of regions of the fruit fly *Drosophila,* and he proposed they were part of a cluster of evolutionarily closely related genes that evolved from one another by duplication and by gaining new functions. He thought that this had occurred within the evolution of flies. Lewis was right about the evolution of the genes, but evolution of these genes was far older than even he imagined.

The break in our thinking came when these homoeotic genes were cloned in the 1980s. Thom Kaufman's lab at Indiana and Walter Gehring's lab in Switzerland discovered that what became known as Hox genes shared a common sequence, the homeodomain, and the more astonishing finding that the homeodomain was found across phyla. The basic key to understanding the evolution of animal development would be recognized from the discovery that many genes that control development are shared all across the phyla. The differences in form have evolved from differences in how, when, and where the regulatory genes are deployed. With the clarity of hindsight I can safely say that we should have realized long before from the branching structure of the animal phylogenetic tree that because all animals shared a single ancestor, they must by common descent share genes that govern development and determine body form. It took a while for that to sink in.

Early in the 1990s, I optimistically sat down to update our book of a decade before. The book I ultimately wrote led me in a new direction. I composed *The Shape of Life* just as the dominance of developmental genetic approaches to evo-devo became locked in. I wanted to stand back from the everyday details of evo-devo and examine what we understood about some of the larger-scale issues. The first part of the book was concerned with the issue of what are body plans, the importance of phylogenetic analyses of gene sequences from across the animal kingdom, and the origins of the animal phyla in the Cambrian radiation. The questions I discussed remain central to understanding how animal life arose. The new data that continually came to us from geochemistry, paleontology and evo-devo have made the metazoan radiation constantly newsworthy.

I was also concerned with the mechanisms by which the structure of animals evolves. I criticized the universal application of heterochrony (evolution by means of changes in relative timing of events in development) in a chapter called "It's Not All Heterochrony." What I intended was to show that heterochrony shouldn't be the default explanation in evo-devo. It is easy to see heterochrony in evo-devo phenomena, and there is a simple reason for that – it is a consequence of how we plot developmental events on a time axis, and any evolutionary change is likely to affect time in some way. We might call these "heterochronies that just fall out because of the way we keep time." That's because any change along a time axis will look like heterochrony but might simply have happened as a consequence of other changes in development, without having been directly selected upon as a timing change.

If we think an evolutionary change is due primarily to a timing change in some process, we have to show it. So my basic criticism was right, but two things have affected my thinking since. The first is that I might have been attacking a straw man as far as developmental genetics-oriented researchers are concerned. Genetic researchers don't seem to do much looking for heterochronies because they are satisfied that evolution runs on changes in gene regulation. I think this is largely correct, but regulation of rates or levels of gene regulation could result in developmentally significant and selectable timing changes in gene action, and that doesn't seem to have been much explored. My thinking has also changed because, in our own work on gene expression changes in larval evolution, we have found immense and important heterochronies are involved. How embarrassing. How interesting. The older school of de Beer and Gould had intuited something important.

A developing animal is not made of just one part, nor is development a single unitary process. Development consists of processes that can be separated from one another in time and place in the flow of development. Evolution consists of changes in features of an animal that have their own distinct developmental patterns and evolutionary histories. I laid out a concept, which I called "modularity," of the role of semi-separate components as crucial to development. Modules can be identifiable as distinct, genetically specified developmental units in a developing organism; for example, the limb buds that give rise to legs are modules distinct from most other developing components of an embryo. In turn, limb buds themselves contain distinct submodules. These modules have particular readouts of genes that are involved in regulating the production of the part of the structure that they give rise to.

From this narrative it sounds as if modules should be sources of developmental constraints that act as a brake to slow down evolution, and at the same time they should offer explanations for how new features can arise. Is that an overly convenient plot device? No, because modules function in distinctly different developmental contexts. Modules that are components of some critical and deeply integrated set of events, say, in brain organization, seem to be conservative. However, modularity means that existing genetically defined units can be "plugged in" somewhere else in the developing organism to allow novel times or locations of expression. That allows the evolutionary co-option of an existing module for a new role in development. The original site of expression of the module can be retained as well. Changes in regulation of their heterochronic timing, changes in strength of interactions, or recruitment into new upstream control elements all make modules important entities that promote the evolution of development. Further, a newly evolved module would not yet be vital to development of central features of a body plan, and thus it could give rise to new elements that would not affect the core of the animal's development. Vertebrates evolved appendages in that way. Vertebrates originally lacked appendages (fins) but acquired them by the incorporation of existing regulatory genes into the evolution of a new modular structure (the developing fin) that was added to and integrated into the existing vertebrate body plan. Thus regulatory genes were conserved in their original function and in duplicated form co-opted in the evolution of a novel module that added an entirely new feature to the vertebrate body plan.

The concept of modules led me to think that the answer to an old but mysterious phenomenon of developmental evolution had a mechanistic answer. That phenomenon is the conservation of common developmental stages among related animals, like vertebrates. This is the observation made by nineteenth-century embryologist Karl Ernst von Baer – that highly distinct vertebrates had a stage part way through development when their body plan appeared, and they closely resembled each other. Earlier and later stages of development diverge, but lizard, mammal, and bird are more nearly identical at this point in their lives than at any other. The stage of greatest similarity among related animals is called the phylotypic stage. In *The Shape of Life,* I drew the progression of development as an hourglass, with divergent early development at the wide top, divergent adults at the wide bottom, and the highly similar phylotypic stages in the constriction of the hourglass. I suggested that the effect came from the maximum cross talk among developmental modules expressing genes characteristic of each developing module at the stage when the conserved body plan is laid out. Denis Duboule envisioned the same hourglass phenomenon as arising from the role of Hox gene organization in their developmental expression in laying out the body axis.

In an amazing coincidence, three research papers by Kalinka and co-workers, by Domazet and Tautz, and by Irie and Kuratani appeared at nearly the same time, in late 2010 and early 2011, to show that there is something special about the developmental genes transcribed during the phylotypic stage. In both flies and vertebrates, genes involved in key developmental processes are expressed and the sets of genes used, the transcriptomes, are most conserved. It seems that a "genomic hourglass" corresponds to the developmental hourglass. I'm delighted that data have confirmed the idea that the hourglass represents a deeply conserved part of the evolution of a developmental body plan, not merely a chance coincidence of events. The results provide a foundation for the classical concept that deep developmental-genetic constraints influence evolution. Thus the evolutionary history of development matters deeply to what we see before us in the living world.

A professor must have a theory as a dog must have fleas.

H. L. Mencken

FIFTEEN

Evolution in the Tasman Sea

WADING IN

In the early 1980s, I started earnestly hunting for the right organism as an experimental system for delving into evo-devo. I thought the ideal animal would be one in which the evolution of early embryonic and larval development could be readily studied because embryos and larvae are crucial stages in development and are simple in cell numbers and types compared to adults. My first efforts were made using the familiar sea urchins of the Northern Hemisphere. I found that we could explore evo-devo at the gene level in sea urchins and published our first evo-devo paper in 1984. In it we showed that a major innovation in the expression of histone genes in sea urchin eggs had taken place with the origins of advanced sea urchins in the Mesozoic, while brontosaurs munched their way across the landscape. We could thus correlate a unique gene regulatory mechanism with a set of macroevolutionary events in sea urchin evolution. But the events were too distant in the past to help unravel ongoing developmental evolution. So I'd have to look farther afield.

In 1985, about the time our molecular phylogeny studies of the animal phyla were starting to produce results from the piles of x-ray films of DNA sequence data we had collected, I gave a seminar at the University of California at Santa Cruz. While there, I talked with John Pierce, a biologist who studied the reproduction of marine invertebrates. I told him that I was looking for a suitable marine embryo system to study the evolution of early development. He handed me a copy of a paper published in 1975 by zoologist Don Anderson at the University of Sydney about an Australian sea urchin with a remarkable embryo. I read it on the flight home. That plane ride was a transformative moment, and it was in a way a cosmic joke.

I had studied the development of the commonly used northern temperate zone sea urchin, and here was something completely discinct, a sea urchin whose development was unimaginably different from that of the sea urchins I knew so well. We all have strong cultural biases about the things we know about nature. So it is with developmental biologists. We "know" that snakes lay eggs, that mammals don't, and that that frogs grow up from tadpoles. Nearly all would say that marine invertebrates produce characteristic larvae distinct from the adults in body plan. Well, all these statements are mostly true. Most snakes do lay eggs, but rattlesnakes and a number of other snakes bear live young, among mammals the furry platypus and the spiny echidna lay eggs, and although the frogs we are familiar with in the northern temperate zone start life as tadpoles, many tropical frogs don't bother. All this was well known to naturalists, who haven't kept it a secret, but not to developmental biologists, who like to think of a canonical form of development for each kind of animal. Unexpected embryo and larval evolution is everywhere if we could just see it. It was only when I began looking for organisms that might have evo-devo potential that I discovered that lots of tropical frogs just skip the tadpole as passé and lay large eggs on land, which develop directly into a little froglet. And then there were similarly deviant sea urchins.

Sea urchins became classics for the study of the mechanisms of fertilization and development in 1877, when Oskar Hertwig saw animal fertilization for the first time by the simple expedient of adding sea urchin sperm to sea urchin eggs – of such simple genius is fame born. In textbooks, the fertilized egg undergoes embryonic cell divisions and develops via a complex feeding larval form called a pluteus. This larva has a bilaterally symmetric body plan, which is vastly different from the pentameral adult. The Southern Hemisphere and the deep seas held a surprise – the pluteus isn't necessary for development of a sea urchin. It's important to understand that marine animals have two basic kinds of development. Direct development is what we see in vertebrates and arthropods, and in some more primitive phyla. In these phyla an embryo produces a chain of larval forms that have the same body plan of the adult and overtime become more adult-like. You might argue that a tadpole doesn't look like a frog or that a maggot doesn't look like a fly. That's true, but the underlying body structures of these larvae are the same as the body plan of the adults.

Something different, though, called "indirect development," is how most living marine animal phyla arrange their growth. In those phyla, the egg produces a larva completely distinct in body plan from the adult. For example, the pluteus is a bilaterally symmetric filter feeder, whereas in its adult form, a sea urchin has five-sided symmetry and feeds by chewing algae off the sea floor. When competent, the larva undergoes a rapid metamorphosis that transforms it into the tiny adult that has been developing within the feeding larva. Phylogenetic studies indicate that early animals likely developed directly, with indirect-developing lineages evolving later in some lineages of animals but not in others. Sea urchins belong to a lineage that evolved feeding larvae, but some species have re-evolved direct-developing embryos. These are not identical to the embryos of ancestral animals, but they bear striking similarities in their properties. These shared features include very large eggs and a lack of larval feeding structures.

Sea urchins were "known" by developmental biologists to develop in the way I was familiar with, via the famous pluteus larva, except when they evolved something else down in the deep seas, the South Pacific, and Antarctica. About 20 percent of sea urchins have dispensed with the pluteus. The *Australian Journal of Zoology* wasn't a journal I had ever read, but in its pages two Sydney biologists, Williams and Anderson, thoroughly described the development of a common sea urchin, *Heliocidaris erythrogramma* – a gift, the ideal organism for what I wanted to do. I hadn't known there were sea urchins that had performed the evolutionary trick of shifting from a pluteus larval mode of development to one where the larva was reduced or lost. I would later learn that *H. erythrogramma,* with its odd larval features, lived spine to spine with another species that had a pluteus – its sister species, *H. tuberculata.* This was a pair of organisms I thought had everything I needed to revolutionize the study of marine embryo evo-devo. They would land me on the golden shores of Australia.

H. erythrogramma development was striking and puzzling because the embryos looked so different from the embryos I knew. Further, I really didn't know much about sea urchin development beyond formation of the pluteus larva. Not a lot of research had been done on the development of the adult within the pluteus as it grew and matured. Biologists studying the molecular biology of early developmental phenomena didn't need to delve further than the early pluteus larval stage. In some cases, their studies didn't go much beyond fertilization. Most sea urchin developmental biologists didn't even consider evolution a relevant issue. I got in touch with Don Anderson in 1985 and asked him about coming over to Sydney University to meet these curious urchins. I itched to try them out as a potential model for the study of evo-devo. Don no longer worked on *H. erythrogramma,* but he invited me to come and have a look. As head of the School of Biological Sciences, he would help me enormously. I landed in Sydney on New Year's Eve, as harebrained as possible a time for arriving. Australians take their holidays seriously. Everything in Sydney was shut down – in the university and in much of the rest of the city. I spent my first couple of days walking around Sydney and waiting out the rain in a pub in the Rocks, the colonial settlement that became Sydney and to this day remains an expensive tourist destination. It's just a short stroll from there to the spectacular Royal Botanic Garden, where I saw my first un-mummified sacred ibis wandering about.

Once the workweek had resumed, members of Don's staff helped initiate me into the mysteries of collecting sea urchins in Australia. We headed out to Captain Cook's original landing place, Kurnell, just inside South Head at the opening of Botany Bay. In April 1770, Captain Cook and his expedition's naturalists, Joseph Banks and Daniel Solander, landed there to meet the continent's unknown fauna and flora. The first kangaroo ever seen close up by Europeans was not shot by Cook's crew until four months later in northern Australia, but Banks and Solander collected plenty of plants unknown to science in one frenzied week. Hence, Botany Bay.

From the place where Cook had stepped out of his longboat onto Australian soil, I looked out onto the rock platform where *Heliocidaris* lived. The light filled everything. Fronds of brown kelp rose and fell in the waves just off shore. A carpet of purple sea urchins covered the ledge of ocher sandstone. Each tennis ball–sized urchin was nestled into a pit it had scraped grain by grain into the rock. I made my first ungainly Southern Hemisphere snorkeling attempts using a blunt-ended diving knife to pop sea urchins out of their pits. This is trickier than it sounds when you are bobbing around in shallow water at the edge of the open ocean. Under windy conditions, surf turns sea urchin habitat into a deadly churn. Under calm conditions, the steady rollers come in from the vast Pacific and you are gently lifted and lowered a few feet by the swell. A dreamy state until your knuckles are deposited onto the pointy spines of a sea urchin or your aim is thrown off and you knife the urchin. It takes a few tries to catch on to collecting in rhythm with the waves. After a bad case of sunburn and several disasters with urchin die-offs in my sea tanks, I finally got *H. erythrogramma* eggs to fertilize and embryos to develop. It proved to be a stressful and exhilarating month.

To my great fortune, I soon met and got help from two experienced divers, Heather and Craig Sowden, technical officers in the school. Craig a short time later moved to the developing Sydney Aquarium, where he would design innovative exhibits. They are two of the best marine collectors and naturalists one could ever hope to know. Heather and Craig became good friends and over the years have taught us an enormous amount about Australian marine life and the natural history of the bush. Each outing with them was always a field course, filled with first-time wonders such as colonial spiders, sea eagles, ghost shrimp, giant sea slugs, the poisonous sea urchin *Toxopneustes,* pigmy lion fish, blue bottle jellyfish, legless lizards, and many others. Because of Heather's intimate knowledge of Sydney marine life, we have been able to carry out a far richer research program with the sea urchins of the area than we ever imagined possible.

THE EMBRYO'S TALE

Once I began studying the embryos of *Heliocidaris erythrogramma* in the lab, I found that I really did not understand what I was seeing in these unusual embryos. Once back home, it would take me a few months of analysis and reading to recognize the profound evolutionary transformations that embryos could undergo in a geologically short time. That in itself was a major insight, because it had been long assumed that early stages of development would be difficult to change in evolution, as all of later development had to follow on what came earlier in the embryo.

The hypothesis required that mutations affecting the early stages of development would be lethal, and so development had to be evolutionarily conservative. One of the first papers we published on the rapid evolution of the *H. erythrogramma* embryo came back with an intriguing reviewer's comment: "You have almost convinced me that embryos can evolve." Almost? That comment was weighted with a load of fading historical and theoretical ideas.

Prior ideas no longer match what we know about early development, or even what the fossil record tells us about how the first animals developed. The theoretical idea is that the first events of embryonic life should be the most conserved because the earliest events of development underpin everything that follows. This view is reinforced by the historical interpretation of Ernst Haeckel that the ancestral animals were similar to living embryos and larvae. He envisioned the evolution of new forms from the addition of new developmental stages to the string of former stages of the ancestor to make new stages of development in the descendant. Neither idea is correct, because both views are based on misinterpretations of what early stages of development represent. In fact, most embryos can accommodate a surprising amount of experimental interference and still regulate their subsequent development. In addition, it is clear now that components of larval and adult development can be dissociated in evolution and result in a new and still viable developmental process.

During my first visit to Sydney, the physical effort of collecting sea urchins and doing all the lab work by myself was enough to make me lose fifteen pounds in four weeks, despite never skimping on beer or hearty eating. The impracticality of working alone soon became clear and made me realize that to keep doing research in Australia I would have to bring collaborators – generally graduate students or postdocs. Doing that led to many years of intense work periods in Sydney with members of my lab. Many of that extended team of students went on to highly successful research careers, two outstanding evolutionary developmental biologists, Greg Wray and Jon Henry, among them. When we worked together in Australia in the late 1980s, they provided the data that let us understand how radically the development of *H. erythrogramma* had evolved and what kinds of additional approaches would be necessary. This early work was important, because there was a trend at the time to view embryos like those of *H. erythrogramma* as simply simplified plutei, just another example of "degenerate" evolution. It turned out to be far from degenerate – unexpected new features had evolved as well as losses. By the late 1990s, Beth was able to join me in Sydney so that we could work together. She designed some of the most penetrating studies we did there, notably the creation of cross-species hybrids and learning what they taught about the robustness of development in the face of genetic perturbation.

The embryos of *Heliocidaris erythrogramma* presented us with five kinds of puzzles. Among sea urchins, the pluteus larva is found in the majority of species and is highly conserved over evolutionary time. But some species have evolved nonfeeding larval forms that develop rapidly and differ greatly from the pluteus, which seems contradictory. How can a larval form be selected to persist for tens of millions of years and then in some species under a different direction of selection rapidly evolve into a different larval form? How fast is such an evolutionary change? What happened? Has the evolution of a new developmental pathway produced anything truly new? What can all this tell us about the evolution of development of the first animals? No one should think that I had all these issues clearly in mind when I began, but the unusual evolutionary history of the organism that I had chosen inevitably raised them.

Developmental biologists seek to ferret out general mechanisms from individual developmental processes. Evolutionary diversity is seen as noise. Variation within a species also was not desirable, because absolutely reproducible experimental results were wanted. Thus a common species that is suitable for lab work would be singled out as "model" species. It may not be at all a universal representative of a group of animals (*Drosophila* is not typical of all insects) or reveal evolutionary variants. Developmental biology got its start in Europe and North America, where differing developmental modes among the creatures of interest often didn't exist. Based on their European or American experience, and because they already had well-studied model species, developmental biologists ignored species like *H. erythrogramma,* which they thought of as unimportant oddities. I was beguiled by the fact that no one studying sea urchin development knew much about those that developed differently than the few species studied in Europe and North America. I saw anomalous creatures from other regions not as deviants but as keys to learning how embryos evolve and keys to reaching back to understand the first animals and the origins of development.

Animals originated in the sea, and so did the processes that came to make up animal development. One of the most striking insights was that immense evolutionary changes in early development happened in a geologically short period of time relative to the roughly 100 million years it took to get from the first sponge to the Cambrian explosion. The time span separating the greatly different developing embryos of the two *Heliocidaris* species is four million years, a blink of the eye in geological time. A time estimate of the degree of divergence between Australian east versus west coast populations of *H. erythrogramma* suggests that three million years is a better maximum estimate for the evolution of the developmental changes of *H. erythrogramma* from its ancestor. Our Sydney University colleague Maria Byrne and her collaborators estimated that similarly astonishing developmental evolution in starfish has taken as little as half a million years. Development evolves with stately slowness – except when it evolves with almost unseemly speed.

The main lessons of our work on the genes involved in the evolution of sea urchin larvae has been finding the surprisingly fast evolution of developmental regulatory systems in *H. erythrogramma.* These systems are composed of evolutionarily conserved genes, but their linkage to other genes that control them can evolve to modify the output in development. The term "upstream gene" is a metaphor used by molecular biologists for genes that operate as the chief controllers of other genes in such networks of genes. We found the upstream genes we studied to be conserved in their basic roles in *H. erythrogramma.* The controlled genes are called "downstream genes," and they encode the proteins that make things happen in the cells of particular tissues. The control of expression of these downstream genes has evolved fluidly.

Developmental biologists Veronica Hinman and Eric Davidson made a crucial comparison of gene control networks across a large evolutionary distance. They compared a gene network controlling cell identities that had been characterized by the Davidson lab in sea urchin embryos to the homologous network in starfish embryos. The core network was conserved. The two organisms they compared were both echinoderms, but they had separated in evolution nearly 500 million years ago. This would seem to show that gene network evolution might be slow and conservative in some components but rapid in more peripheral components. In *H. erythrogramma* development, compared to that of its sister species *H. tuberculata,* we see changes in downstream gene regulation that have produced a new kind of larva, but in four million years or less.

What we discovered is that an enormous sum of genetic change took place within a conserved regulatory background. *H. erythrogamma* larvae have no mouth. They evolved from a species that has a feeding pluteus larva with a mouth. The same upstream gene regulatory system needed to set up the oral side of the embryo is at work, but *H. erythrogramma* expresses some downstream control genes in different patterns than the ancestral pluteus. We were able to identify and show the role of one of the genes that was crucial to the loss of a mouth in the evolution of the nonfeeding *H. erythrogramma* larva. Injection of the messenger RNA of the gene and expression in *H. erythrogramma* resulted in re-formation of the mouth lost in evolution. Genetic changes in the evolution of *H. erythrogramma* development include more than changes in amount or place of expression. Timing changes in gene expression are prominent as well. The old classic, heterochrony, is still a living part of evo-devo but now can be studied as a genetic mechanism.

We asked if cross-species hybrids, which Beth made by fertilizing *H. erythrogramma* eggs with *H. tuberculata* sperm, could reveal how the control systems of these two divergent species could interact when reunited into a single embryo. We expected to see at best a few hybrid features. When we anxiously watched our first culture of hybrids, we found that they developed just fine and metamorphosed. However, the hybrid larva weren't really like either parental species. They looked like the more primitive larvae of starfish. These hybrids let us understand how robust developing systems can be to genetic perturbation. Mutations in early development might transform the path of development, but the show could go on. When we made hybrids between *H. erythrogramma* and distantly related species that had evolved direct development independently of our species, we discovered that they had evolved by similar genetic routes.

Our cross-species hybrids were immensely important because they demolished the seemingly compelling idea that early stages of development could not evolve easily because early stages would be especially sensitive to any mutational perturbation. Forcing genomes of the two species into the cytoplasm of the egg of the maternal species and achieving harmonious development seems like pushing the system to the extreme. The two genomes in question inform the development of two distinct embryos, larvae and life histories. Yet the hybrids challenged fate and developed quite nicely. Both genomes were expressed as messenger RNAs, and distinctly novel larvae that shared features of both parental species were formed, but in unexpected combinations. Somehow the underlying gene networks could interact to produce a pattern of development different from the parent species but still coherent and able to undergo metamorphosis into a little sea urchin. But only with the right egg.

One striking feature of the hybrids is that success happened in only one direction. We did the reverse cross of *H. tuberculata* egg fertilized by *H. erythrogramma* sperm. The embryos died young. The same two genomes were combined in both kinds of hybrid, but only hybrids in which the *H. erythrogramma* egg was fertilized by *H. tuberculata* sperm developed successfully. The sex of the parents mattered, meaning that more than just the mixture of genes mattered. This result showed that evolution re-organized the structure of the *H. erythrogramma* egg in a major way. This outcome was consistent with our earlier discovery that major heterochronies had occurred in *H. erythrogramma* so that the first steps of generation of the body axes during development shifted from the embryo earlier in time to the egg. We don't yet know what controls changed the timing.

AN EVOLUTIONARY INTERMEDIATE

Biologists dream of discovering an evolutionary intermediate. If the highly derived development of *H. erythrogramma* evolved from the development of a pluteus, can we find any living sea urchins with an evolutionarily intermediate larval form? It's a truism in court that lawyers don't ask questions for which they don't already know the answer. It's so in this story as well. There are a few species of sea urchins whose developmental pattern is somewhere between the ancestral full-feeding pluteus larva and the direct-developing forms such as *H. erythrogramma.*

One such species lives in Panama, on the yellow coral sands in the warm, clear waters off the Smithsonian Marine Station at Bocas del Toro on the Caribbean coast. We visited our research colleagues Harilaos (Haris) Lessios and Kirk Zigler at the Smithsonian Tropical Research Institute in Panama City and flew from there to Bocas to collect what we hoped would be a sea urchin species intermediate in the evolution of development that could tell us how the shift occurred.

Haris steered the dive boat out from under the thatched roof boathouse at the end of the Bocas dock. The air was warm and damp, and as always, dramatic masses of cumulous clouds hovered on the horizon. We headed for clumps of mangrove trees that appeared to float on the seawater. In a sense they did. This mangrove sea was a unique environment. The clear water was only a few feet deep. Red sponges and yellowish fingers of fire coral adorned the mangrove roots, and like artful robots, white-striped spider crabs scuttled over the coral. We snorkeled in the bathtub-warm water over the pale coral sand bottom upon which foot-long sea cucumbers grazed accompanied by the sea biscuit urchin, *C. rosaceus.* This urchin is hemispherical and large enough to cover a man's hand. Tiny blunt spines carpet the animal, endowing it with the look and feel of a baker's terrible blunder, a hard pastry in need of a shave. Once down to business, we collected biscuit urchins of two species to take back to the marine station, from which we would shed eggs and sperm. Fertilization was simple as was culturing the embryos, which we could examine in Bloomington to let us see what developmental features might be evolving.

C. rosaceus has a larva that looks like the ancestral slow-developing pluteus form typical of the sea urchins beloved of northern biologists, but it develops rapidly, and although it can feed, it doesn't need to. When we examined the insides of the larvae, we discovered that a crucial feature, the coelom, needed to start adult development, was enormously accelerated in its time of formation when compared to the development of the coelom in typical plutei. Development of the coelom in *H. erythrogramma* is even more accelerated. The eggs of *H. erythrogramma* are novel devices that produce juvenile sea urchins from eggs as fast as possible without troubling with pluteus features. *C. rosaceus* is not a direct evolutionary intermediate in the *H. erythrogramma* lineage, but it serves as a wonderful example of what developmental features that the lost evolutionary intermediate form may have had. In *C. rosaceus,* new features are gained before the old are lost, which tells us that the need for speeding up development may have provided the selection for early steps in evolution of nonfeeding larvae.

As for Bocas, it was a charming town with a couple of excellent restaurants that we sampled well. Going to the marine station involved following the paved main street of town, which ended suddenly in a muddy slough through which we lurched in order to turn hard left onto the dirt road that led to the bay. As we left town we noticed a low, cream-colored building prominently labeled in large black letters: "Regional Hospital." Next to it, behind a cyclone fence, was a low, white building labeled "Morgue." Just a few yards farther down the road, a low wall surrounded the neighboring lot. The arched gateway topped by a cross announced the town cemetery. I told Beth that if I got sick in Bocas, she wasn't to take me to that hospital. The odds didn't look promising, but maybe it was just an unfortunate piece of town planning that all three buildings were placed together. Certainly the planning or lack of it at the Bocas airport was extraordinary. The runway comes straight in from the sea. At the end of the paved runway is grass. The grass merges into the playing field of a soccer stadium complete with bleachers. The players happily do their fancy footwork yards from the end of the runway, no fence intervening. While we waited to fly out of Bocas, we stepped out of the airport waiting room to watch the soccer players on the grass in front of us. We were shooed back inside for unclear but firm reasons of airport security, or maybe because the hometown team was losing.

Our visit to Panama let us find our evolutionary intermediate, and it also would let us see more of the tropical rain forest than we had seen in 1969. The Mexican rain forests we had visited in the 1960s were not as dense or rich in animal life as those farther south. In Mexico, we never saw or heard monkeys, which seem to have been heavily hunted. We visited the Smithsonian's Barro Colorado lab, located on an island that was originally a hill around which the rain forest was excavated to make the Panama Canal. When the water was allowed to pool to make Gatun Lake between the locks, the hilltop was isolated. The island is restricted and has only the Smithsonian research station on it. To get there one has to get up early and, in the cold morning mist, catch the dawn ferry from Gamboa dock. The lake is part of the canal and full of the traffic of large ships, which at night appear as strange brilliantly lit entities moving silently on the water. The water is brown from erosion of the mud banks, and huge dredges are kept going day and night removing silt from the ship channels. They pump the mud through massive pipes floating on barges and dump it right back into the lake, so I imagine this profession has an endless future. Near the shore, there are places with water lilies and jacanas, birds with disproportionately large feet that walk on those lily pads. Swimming is discouraged because large caimans lurk in the shallows. Like other crocodilians, they are well equipped to take a bite out of the unwary. Bits of meat from lunch tossed into the shallows bring them forth, agreeably and with snapping jaws.

Although we visited rain-forest sites in Mexico, animal life beyond insects was scarce. Barro Colorado as a protected reserve was another matter. There life abounded, including a number of unusual mammals. The island is only a few square miles in extent but holds more than thirteen hundred plant species, including about three hundred species of trees. There was the sloth that hung out by the field station, agouties everywhere, rare tree anteaters, and howler monkeys. Beth, Kirk Zigler, and I hiked through the heat and humidity, dwarfed by the buttressed trunks of great trees. Vines endeavored to become one with the trees they climbed. A dead leaf lying on the trail opened its wings and transformed into a butterfly. A pair of keel-billed toucans with their outsized bills and bright yellow breast feathers carried on in the deep green foliage. Howler monkeys called from far off and then closer. We found a place on the trail that would intersect the monkeys' treetop highway, and soon the troop came up. They stopped high over us and looked disapprovingly at the earth-bound primates. The males tried to pee on us and threw dung – primate-to-primate stuff. The throwers were all show with no accuracy. Then the troop moved on. We actually were grateful for the dung shower for what it would show us about the rain forest's floor environment. As soon as the dung landed, dung beetles were on it, frantically claiming the treasure, which they quickly rolled into balls for their larvae to feed on. This was a spectacular demonstration of how rare food is on the rain-forest floor and how finely tuned organisms are to find key resources, however modest. There are even species of dung beetles that cling to the fur at the back ends of monkeys to be first in line for any precious dung that might appear. The forest floor is the kingdom of the leaf cutter ants. Long processions of these inveterate underground fungus gardeners crossed open ground. Each ant carried in its mandibles neatly scissored off parts of leaves. In one case, each ant in the stream carried a bright pink flower petal.

By taking a ride around the island with a wildlife conservation officer on his patrol boat, we saw the outer surface of the forest, which is made of the closed upper canopies of the giant trees. The crowns of some of the common trees were brilliant with yellow or purple flowers and easily seen from the water. The conservation officer was a passionate naturalist and knew a lot of the natural history, but our Spanish was not up to his narrative, except for some common names. Poaching is a serious problem, and the wildlife officers are well armed. There are still ocelots on the island. They live on the conveniently sized agoutis (six pounds of goodness). The jaguars, though, are gone, as the island is too small to support such a large cat and on the mainland they have been severely poached. The island is probably the most intensively studied patch of rain forest in the world. However, the rare and elusive still happens. A stray jaguar was photographed on Barro Colorado in 2009.

APPLYING EMBRYOS TO EVOLUTION

How do we apply our observations of the remarkable evolution of major features of embryos and larvae to understand the explosive metazoan radiation at the start of the Cambrian? We realize that embryos arose with the origins of animals, but only later in the Cambrian did larvae evolve to swim and feed in the sea.

H. erythrogramma shows that larval evolution is fast, and by analogy it is likely that early animals and their embryos also evolved rapidly. Relatively simple gene regulatory networks would have allowed rapid evolution of embryo forms. Further, the small, simple body plans of the first animals would have tolerated, as minor, changes to features, such as axes of symmetry, that would be major or impossible changes in their complex descendants. It's not only our larvae that support such notions. The fossil record suggests that all the basic animal body plans arose in a few tens of millions of years, probably in no more time than separates us from monkeys. Even the eye, so often used by creationists as an example of a structure too complex to evolve, turns out to be a powerful example of how rapidly novel structures evolved at the dawn of the animal kingdom. Nilsson and Pelger did an illuminating computer simulation of the steps of eye evolution that showed that an eye like ours, with a lens capable of forming an image, could have evolved from a simple light-sensing patch of tissue in a small fraction of the time of the Cambrian radiation.

If, as phylogeny indicates, the first animals developed directly, what do all those diverse feeding marine larvae represent in evolutionary history? They seem to have arisen as an evolution of a secondary but quite distinct stage of animal body plan evolution by animal phyla during the Cambrian. The fossil record of embryos and larvae supports the hypothesis that early in animal evolution, eggs and embryos were large and larvae nonfeeding. By the end of the Cambrian, the fossils show clearly that small mollusk larvae have replaced large, more primitive direct developing forms. We don't know if this story applies as widely to other phyla as it does to echinoderms and mollusks, but the body plans of at least some animal phyla had two bursts of evolution. In the first, just before the start of the Cambrian, the basal adult body plans evolved. In the second, some millions of years later, indirect development, distinct larval body plans, and metamorphosis evolved. Although we mostly think of adult evolution as the important formative part of evolution, remember that most of the early animals were not very motile as adults. Some could crawl and a few were swimming carnivores, but many were attached to the sea floor for life and fed on plankton wafting by. The evolution of novel larval body plans would strongly affect such important traits as the dispersal of young to new sites where they could settle, separation of larval from adult feeding needs, and escape of larvae from predators.

Don't ask if it's poisonous. Ask if it's deadly.

Australian bush advice

SIXTEEN

An Alternate Present

AMONG THE LIVING FOSSILS

I was drawn to Australia because of the extraordinary possibilities it offered to study evo-devo in a marine embryo that had evolved with abandon. Everyone knows about kangaroos and a few other oddities such as giant fruit bats, photogenic koalas, and good beer. So it was for me when I first went Down Under. But as a biologist, I was soon overwhelmed by the endless strangeness of Australian life forms. I still am, as a matter of fact. What many return visits have taught me is that the weirdness is more than koala deep. Australia is not lost in time; it is an alternate, living evolutionary outcome on planet Earth resulting from millions of years of evolution in isolation. Australia's last direct links to the rest of the world started being ruptured by inexorable plate tectonics about 80 million years ago, when dinosaurs still roamed Antarctica and Australia. A connection to Antarctica and from Antarctica to South America persisted until about 40 million years ago and then was severed with the final drift of Australia to the north. The ancient plants of Australia were derived from a forest that extended from South America through South Africa and Antarctica to Australia, and the botany of that long-dispersed Gondwana forest has left its mark. The Antarctic beech that long ago grew in the forested vales of Antarctica still grows in the cool highlands of Australia. The exotic banksias and grevilleas that bloom in Australia are the evolutionary cousins of the South African proteas beloved of flower arrangers. The state flower of New South Wales is the gorgeous red waratah, which has close relatives in South America.

As Australia dried during the last several million years, the ancient rain-forest ancestors of the myrtle family produced a radiation of hundreds of species of fragrant fire- and drought-tolerant eucalyptus trees. The penchant of Australia for preserving living fossils has created anachronistic fantasy forests that combine ancient and newly evolved plants. The southern coastal woods of New South Wales are topped by spotted gum, immense smooth-bark eucalyptus canopy trees, and a scattering of much smaller banksias beneath. On the forest floor, Mesozoic cycads mingle among modern gum trees – cycads everywhere, amazing to North American eyes. That ground cover is composed of eight-foot-high *Macrozamia* cycads that in summer produce two-foot-long seed cones. These are Mesozoic plants that would have been at home in the ancient conifer forests, systematically munched on by long-necked brontosaurs. Paleozoic tree ferns with twelve-foot-tall trunks topped by elegant brilliant green fronds grace low, wet places.

When Gondwana fragmented, Africa split off first, and then, about 40 million years ago, South America, Antarctica, and Australia split from each other, but not before the ancestors of the present animal fauna had entered the soon-to-be-isolated Australia via a still-mild Antarctica. That ancient connection still shows in the fact that only South America and Australia have had such rich radiations of marsupials. These are animals like opossums or kangaroos that carry their young to term in pouches. Everyone is familiar with the Australian marsupials because they have the simple advantages of being large, still living, and heavily promoted by tourism boards. Most living South American marsupials are small "possums" and don't rise to the postcard charisma of koalas, kangaroos, and the feisty Tasmanian devils. However, some of the extinct South American marsupials were spectacular. If the lion-sized South American marsupial saber-toothed "tiger" *Thylacosmilus* were still among the living, it would get plenty of respect.

The marsupials of Australia are more than big-footed hopping oddities. They represent an entire mammalian ecosystem analogous to what we are used to in the familiar continents, where placental mammals have been the norm for 60 million years. A fossil platypus lived in South America about the time the non-avian dinosaurs became extinct. The egg-laying mammals, platypus and echidna, represent the last of the egg-laying mammals that were advanced mammals in the dinosaur's world but live on as relicts in the lost continent. Marsupials too originated from a single marsupial lineage in South American. Although there were primitive placental mammals in Australia during the age of dinosaurs, they died out by about 50 million years ago to be replaced by marsupials. That seems all wrong if placental mammals are inherently superior to marsupials, but evolutionary outcomes depend on conditions.

Marsupials evidently were superior in the sparse environment of Australia. Kangaroos have evolved a completely novel trick in their reproduction. A female kangaroo can decide on the initiation or pausing of development of an embryo depending on circumstances. Marsupials have evolved to fill the roles of placental mammals in the rest of the world – in their own odd ways. Kangaroos fill the niche of deer and antelope, but they are the only large animals to ever evolve hopping as a way of achieving high speed. Hopping turns out to be a highly energy-efficient means of transport. Kangaroos also have evolved foregut fermentation convergently, by an independent evolutionary path from that which led to unrelated ruminants such as deer. Foregut fermentation allows the cellulose of grass and leaves, which is normally indigestible by mammals, to be processed by symbionts. These can then be digested. The symbionts are bacteria and protists in deer and cows, but they include writhing nematode worms in the ever-original kangaroos.

The fossil record of marsupials, related to kangaroos, wombats, koalas, Tasmanian devils and others, goes back at least 20 million years and documents that living Australian marsupials evolved within Australia from earlier migrants. More recent Ice Age giants included the immense short-faced kangaroo, taller than a professional basketball player; a tree-climbing marsupial lion with flesh-shearing teeth that still haunted ancient Australia when the ancestors of the Aborigines arrived; and a rhino-sized creature called *Diprotodon* (essentially a hugely inflated wombat) that dwelt by the shores of inland lakes before the drying of the continent turned those lakes into immense salt flats. There was even a carnivorous kangaroo – something akin to speaking of a killer bunny. Contingency and natural selection were at work.

The results of the long isolate stream of Australian evolution shattered Eurocentric concepts of normal. The southeast coast New South

Wales eucalyptus forest is bizarre to northern eyes. Man-high dun-colored termite mounds dot miles of forest floor. The mounds, each patiently built up of tons of termite feces, tell just who controls the recycling of plant carbon. Despite their metabolic power, the mounds brood in silence, but in summer these cycad-gum forests are not silent. The woods echo to the territorial calls of male bellbirds, petit but hyper-aggressive olive green honeyeaters. The air rings with their insistent one-note songs, a raucous competition among thousands of wind chimes, each announcing its superior DNA. Male whipbirds call out like singing voices mimicking the crack of a whip. The females reply with an almost instantaneous "weep-weep." Shy echidnas shuffle through the underbrush.

Then there are all those reptiles. Australia has an impressive evolutionary radiation of poison snakes in the cobra family, and their place in the hierarchy of world's most poisonous is a point of national pride. At a station in Tasmania where tiger snake venom is harvested for the preparation of antivenin, we talked to the man who "milks" five-foot tiger snakes for venom by gently massaging the poison glands and collecting the precious drops from the fangs. He said wistfully that tiger snakes are only the fourth most dangerous of the world's snakes. He felt that their venom was top notch, but with a more efficient delivery system (they make do with fangs only a few millimeters long), "they could do better." Why would you want them to do better if you handle them every day? Or even if you don't? Beth and I later encountered a full-grown tiger snake on a Tasmanian hill trail where we had gone to look at *Anaspides,* a living fossil crustacean from the coal age that still survives in mountain streams. As we walked down, we ran into the snake sunning himself in the middle of the path. Snakes are deaf but sensitive to vibrations. At a safe distance we jumped up and down to let him know by the vibrations that he should leave for a quieter spot. He coiled into striking position. Prodded by my Y chromosome, I poked him gently with my long walking stick. He lunged. We jumped back, conceded the point, and took another trail down. Beth is still mad (at me, not at the snake).

Snakes are what everyone talks about, but the other reptiles are unexpected spectaculars of evolutionary history. Australia's turtles share with those of South America an altogether ungainly and strange mechanism for retracting their heads into their shells. These turtles, relicts of the ancient South American–Antarctic–Australian continent, can't retract their heads gracefully deep into their shells like the advanced turtles that evolved in the Northern Hemisphere. Rather, side-necked turtles must lay their heads to the side along the front of their shells instead of pulling them back completely into the shell. It seems a method much less safe from carnivores, but side-neck turtles have survived the test of selection. Their solution is good enough.

Unexpectedly, it would seem that lizards rule. The drying of the continent allowed the evolution of diverse and utterly strange lizards. The lizard hoard dominates the diversity of the dry continent because lizards operate more efficiently in a meager, hot environment than do mammals. The smaller lizards of the desert exploit termites that convert the poorly nutritional drought-resistant plants into high-quality insect prey. The largest lizards are the single-mindedly carnivorous monitors, colloquially called "goannas," which can reach two meters in length. I once was out in the bush north of Sydney when two active and hungry six-foot goannas joined us for lunch. They took the pieces of chicken we offered and bolted them down happily. We kept a wary eye on their claws, which would do an eagle proud. Until the Pleistocene extinctions about forty-five thousand years ago, Australia's biggest land carnivore was *Megalania,* a monitor lizard reaching about fifteen feet in length and weighing a few hundred pounds, about the size of a substantial crocodile.

For my money, the most winsome of Sydney lizards is a species of giant skink nearly as long as your arm but with comically short legs. In their mouths is a large bright blue tongue, hence their affectionate common name, "blue tongue." These lizards rapidly get used to being handled, and they enjoy escargot. A large garden snail is a blue tongue treat. They crack the shell with their teeth and, using those lurid muscular tongues, carefully peal it from the living snail, which they then swallow whole.

WHAT GOES ON INSIDE FALLEN LOGS?

The evolutionary wonders of Australia also include the lowly invertebrates. In 1986 we were deep into our study of the molecular phylogeny of the animal phyla, and I had a shopping list of creatures not available at home whose RNAs could be revealing of their evolutionary histories. Scorpions and peripatuses were high on my wish list. In truth, though, no one likes to be a novice collector of scorpions. Peripatus is harmless, but it is a special and rare creature that I had no idea how to find. Peripatuses are onychophorans, the only living descendants of the enigmatic lobopod animals that had once dominated Cambrian oceans. These elusive animals share a close ancestry with arthropods such as lobsters, spiders, and insects and so hold vital clues to the origins of arthropods. I asked around for someone with the necessary experience and love of invertebrates. I wanted to ask his help in collecting invertebrate taxa from which we could extract ribosomal RNA (the form of RNA that is part of the ribosomes that cells use to make proteins) for our molecular phylogeny project. Everyone gave me the same name: Noel Tait.

I first met Noel in January 1986. Noel was then at Macquarie University just across Sydney. I called him up and he invited me over to the cluttered office he shared on a walk-in basis with a cheeky black-and-white bird called a peewee. Noel is one of the great characters of Australian zoology. Given the number of outsized characters celebrated in Australia, I mean this as rare praise. Irreverent, fast-talking, wiry, and infatuated with anything that moves in the bush, Noel is to my mind the archetypical Australian naturalist. He can find anything. When I asked for scorpions, he showed up with a jar of them, which he casually dropped off for me (chunky gray *Urodacus* with powerful sting-tipped tails). I had to figure out how to convert my gift of touchy scorpions into RNA without getting stung. Long steel forceps were a handy accessory. Once getting past putting them into their final sleep by anesthesia with CO_2, I discovered as I dissected that some were gravid females. Developing baby scorpions were stacked up in the body cavity connected to a kind of placenta and would have come into the world by live birth.

Although the Cambrian lobopods were marine, the living onychophorans now live on land, and only in the Southern Hemisphere. It had been hypothesized that onycophorans were a "missing link" between annelids and arthropods in phylogeny – a tenacious idea that would have to be removed from the minds of zoologists with molecular pliers. Peripatuses are close relatives of the ancestors of arthropods. They have fifteen pairs of unsegmented legs and so have the look of an *Alice in*

Wonderland caterpillar. Peripatuses are often called velvet worms – velvet because they are covered with a silky pelt, and worms because almost any elongate invertebrate smaller than a kitten gets called a worm.

These enigmatic creatures live in rotting logs. Odd and seemingly defenseless, they defend themselves effectively from ants and biologist's forceps by accurate squirts of superglue from giant glands that feed powerful nozzles on either side of the head. With their binocular vision, they are good shots, and their velvet coats allow them to walk through tangles of their powerful glue without any self-inflicted embarrassment. The females of the Australian species bear live young and seem to be devoted mothers, that is, they don't eat their own babies. Peripatus males have sexual practices that put them high on the list of weirdest and least romantic lovers in the animal kingdom. They have a pit in their forehead that has sharp little spines in it. When the mood strikes, they transfer a packet of sperm to the forehead pit and charge off through their tunnels to find a female. What happens next depends on the species. On finding their true love, the males of some species just stick a sperm packet anywhere on the outside of the female's body. The sperm pass through her integument and find their way through her body cavity to her gonads. In other species, a more cerebral sort of mating takes place in which the males place the tops of their heads on the female's genital opening and drop off the sperm packet. Peripatus sex has been described as "casual." Child custody issues don't arise.

It turned out that Noel had been inadvertently lured into studying peripatus. For a century, the peripatuses of Australia had been all lumped into a single species called *Euperipatoides leuckarti.* Noel taught invertebrate zoology and collected a wealth of invertebrates for his teaching labs. He varied his collecting spots in the Blue Mountains and noticed that *Euperipatoides* from one mountain could look different in color from the supposedly identical ones on the next ridge. He didn't just shrug off this seemingly trivial observation. A look by scanning electron microscopy showed that those sperm pits mounted on the heads of males differed greatly. Molecular studies by Noel and his colleague David Briscoe provided the shocker. Some of the distinct populations along the mountain range were millions of years separated from each other, while others were very close in time. Ultimately the original single genus and species was recognized to consist of a dozen or so genera and perhaps fifty species. Living fossils peripatuses may be, but they continue to evolve at a merry rate. Noel became a sort of guru and practical guide in the bush to many of us Northern Hemisphere biologists and paleontologists who needed to collect "velvet worms" for a variety of studies of early animal evolution and the Cambrian radiation. Our molecular phylogenetic studies and work by other colleagues on gene expression in peripatus embryos helped illuminate Cambrian animals and the evolution of arthropods.

I've had the pleasure of taking numerous trips with Noel into the gum forests of the Blue Mountains to seek out the elusive peripatus from their lairs in giant decaying logs that bask peacefully in the dappled sun of ferny forest glades. Peripatuses live an ample life in their rotting logs, eating smaller invertebrates like termites. We disturbed their peace using oyster knives with wimpy three-inch blades, which was slow work but meant not missing or destroying critters, and it developed strong fingers. These trips were a quick introduction to the invertebrate woodland fauna that share the logs with peripatus, the pale termites, giant black neotenic roaches, lurid yellow-and-green-striped land planarian flatworms, and (if you have different color preferences) designer pink and blue spiders. I also met up with a great cast of variously toxic and belligerent arthropod characters, scorpions, funnel web spiders with hostile attitudes and the outsized fangs to go with them, feisty bull ants happy to take on anyone, however large, and bestow a fiery sting, rippling centipedes, and insidious ticks. On my first trip, we had the company of a few national park rangers, who looked at our permits and stayed around to see the scientists at work. In reality they were fascinated at the sight of people crazy enough to poke around in funnel web spider logs with their fingers. "Better you than me mate" was their cheery parting.

During a break from digging, I walked over the glade to look at the fall of gum tree logs, some as much as five feet in diameter and a hundred feet long. Tangles of these logs are hard to walk around, so one climbs over. I stood on one of these big horizontal tree trunks, master of all I surveyed, until the log turned out to be hollow and its roof collapsed. As I dropped in slow motion amid the splintering of rotten wood into the dark cavity, I thought, Oh God, what's in here? imagining a cobweb-filled gloom full of excitable red-bellied black snakes and funnel web spiders. Only me, it turned out. The log was an immense, thin-walled hollow tube with a clean, long cavity, smooth and uninhabited. I climbed out, chastened. Our collecting jaunts attracted an infinite supply of annoying but harmless bush flies that charmingly wanted to crawl into our eyes to drink. But there was more joy in the underbrush. The most sinister uninvited guests were the land leeches, which are even worse than they sound. Rubbery and tan, about two inches long, head held high, they rustle through the leaves, moving ever closer, then, unfelt, they creep into socks and pant legs to seek armpit or crotch. Once there, they painlessly slit the skin and gorge on blood. The end of the day brings scientists together, standing around the car searching all those private places for unwelcome guests.

Going into the bush with Noel was not just a scientific experience. He knew how to find the iconic parts of natural history. The rock platforms of the coast of New South Wales near the town of Ulladulla preserve a seabed that lay under floating glacial ice in the Permian Ice Age, 270 million years ago. The melting of the floating ice dropped polished stones ground by glaciers into the mud below, and now these are preserved as evidence in the rock. The fossils of giant brachiopods big enough to cover a hand attest to the slow growth of animals in the cold water. The drift of Australia north, starting about 60 million years ago, would carry its evolving fauna away from the icebox that is now Antarctica. On a trip to look at the fossils, we stopped at Pebbly Beach, where Robyn Stutchbury, Noel's wife, who had worked there as a geologist, showed me the rock platform famous for fossils of U-shaped worm burrows and other traces left by once-living animals in the ancient seafloor. Pebbly Beach is where I saw my first lyre bird dash across the track after I'd heard them sing in the bush all day, mimicking the sounds of whip birds and, oddly, even the sounds of trucks in low gear and chain saws. I also met my first satin bower bird and looked admiringly at his bower, which was carefully built of twigs and piled with blue plastic trinkets spread out to attract his lady's favor. His plumage was indigo. His eyes were blue glass. We have a good idea of what color she likes.

These outings weren't just gustatory occasions for happy leeches. Noel would always bring a hamper of food for the expedition. The first time he did so, I thought, Okay, it's Australia, vegemite sandwiches and beer. But no, Noel would produce an entire gourmet meal of roast chicken, homemade rolls, butter, and elegantly dressed salad, complete with an array of excellent wines. He prepared his repasts late into the night the day before and served them in the bush after our encounters with the log creatures, which could include the painful bull ant sting for anyone careless enough to put a finger within range. Lunch was set on a folding table. We would sit around in the dappled shade of the gum trees waving off the flies and feasting. On one trip he introduced us to lamingtons, traditional (but not likely to be a splashy hit in world cuisine) Australian confections made by putting chocolate and shredded coconut on pieces of two-day-old white cake to give it a second life. On the trip in question, we had a full car, Noel driving, with marine biologist Greg Rouse (a guru of marine worms), zoologist Lennart Olsson (visiting from Germany to study the Australian lungfish with Macquarie University professor Jean Joss, who has the rare ability to do evo-devo with them), and Beth and me, all well packed in. We left early in the morning with the promise of breakfast along the way.

As we drove west from Sydney into the mountains, I spotted a restaurant that advertised breakfast. "No worries – there will be a better spot up ahead," said Noel without even a glance. Then we passed another roadside restaurant nestled in the lush tree ferns. I pointed hopefully. "Not yet." Noel dismissed it as I hungrily watched the eatery recede into the distance. Finally, once well into the mountains, Noel pulled onto a dirt track leading to a finger of bush surrounded by spectacular one-thousand-foot sandstone cliffs falling away into the rising mist. "Here's the spot," he announced, and then, as we stood in the chill air, a little huddle of biologists, he pulled out lamingtons and coffee from his hamper. The coffee steamed in the nippy air. It was a Noel moment, a treat of Australian bush food legend for his hapless visitors. Beth wandered off with her coffee and a dread lamington. She came back with a report that some of the local snakes were warming up in a patch of sunshine. It was a pair of young death adders, and they gave the morning a magical feeling. No one poked them.

It may seem dreamlike to hunt peripatus in the Australian hills, but they are, after all, Cambrian animals, living fossils. They represent a kind of evolutionist's fantasy, almost like seeing a live trilobite from the view port of a deepwater submersible. Now, imagine being offered a wager on the existence of fossil embryos from the time of animal origins. No one would have thought of betting on that proposition. Just think how exciting and informative it would be if we could see directly how some of the first animals developed. What questions we could ask.

The earth is not a mere fragment of dead history .

Henry David Thoreau

SEVENTEEN

Biology Meets Fossils

ANIMALS APPEAR

I've never lost my interest in fossils or my love of tracking them down in their rocky haunts. Sometimes that haunt lies in the splash at the foot of seaside cliffs. Standing there, I'm lost in time, the fossils merging with the heave of a sea in which an imagined swarm of alert, large-eyed ammonites still float with colored tentacles projecting from the mouths of their graceful chambered shells. Disentangled from fantasy, fossils are priceless tools for seeing through time in order to answer questions in evolution. In particular, I've wanted to see far back beyond those beautiful ammonites to the origins of multicellular animals, the organisms that include our ancestors.

In the past decade there has been a revolution in discoveries of Precambrian and Cambrian animal fossils; a whole half-billion-year-old zoology is appearing before our eyes and thought. These fossils are extraordinary enough, but even more amazingly, fossil embryos from the dawn of animal life have turned up as well. The flash in the fossil record that marks the origin of advanced animals has been variously called the "Cambrian explosion" and the "metazoan radiation," names that come from the observation that the animal body plans found in the animals that live today appear with relative suddenness in the fossil record, in strata deposited during the Early to Middle Cambrian time. The official start of the Cambrian was 542 million years ago, and this fabled geological period closed up shop 488 million years ago. The "explosion" had ended by about 520 million years ago. In fact, the metazoan radiation had begun slowly 100 million years before that and reached a crescendo in the Early Cambrian.

Despite all the data that has come from the genomes of living organisms, the fossil record is our chief direct link with the deep past of life on Earth. The simplest definition of a fossil is that it is the remains of some extinct creature. The value of fossils comes from the information they provide on the way extinct organisms were built and worked as living creatures, and on the processes and steps of evolution. Fossils are dated by geological techniques that give us their placement in the sequence of life and even their age in years. Fossils give us the ruler by which we can calculate rates of evolution. We can see degrees of relationships in gene sequences, but ultimately, to determine the timing of evolutionary events, we have to go back to geological data. Correlation of evolutionary events recorded by fossils with the dating of rocks by use of radioisotopes is attaining ever-greater precision. The dates I scatter about so freely in these pages come from these well-tested sources.

The evolutionary information that a fossil can give us depends on how well we can place its relationships to living forms and how much information a fossil still preserves. Both of these kinds of information are likely to slip into ambiguity the deeper into the past we look. The reasons are simple. Relationships become obscured because the farther back we go, the fossils will look ever less like living lineages. A mammoth is different from a living elephant, but the close relationship and great degree of similarity to living Indian elephants lets us easily understand the ancient animal. The principle of homology allows identification of features present in both fossil and living forms, even if the homologs are not identical. Exploitation of homology has allowed, for example, paleontologists to trace the evolution of our inner ear bone from parts of the ancestral reptilian jaw during the evolution of early mammals. But when we look for the common ancestor of butterflies and bears, the going gets tougher. For very old fossils, the similarities grow fainter, until we reach the enigmatic Ediacaran animals and find it very difficult to know just what they were and how they functioned.

Before the start of the Cambrian, only a few animals have left us their fossil remains. Within a few million years of the 550-million-year-old Ediacara animals, a whole marine aquarium's worth of spectacular animals emerged as concrete forms in the fossil record by the time of the Chengjiang animals, 520 million years ago. This apparently sudden origin of advanced animals has long cheered creationists, who seem to forget that tens of million years is actually not equal to instantaneous creation in a twenty-four-hour day. Still, the apparently abrupt rise of animals troubled Darwin and generations of paleontologists who followed. The discovery of even better Cambrian fossils in the Burgess Shale of the Canadian Rockies in the early twentieth century only sharpened the difficulty, because these fossils show a diversity and complexity in Middle Cambrian animal life that rivals later times. The preservation of kinds of feeding structures and even prey items in fossilized guts suggests that the Cambrian had food webs a lot like those of modern ecosystems. More recently discovered and exquisitely preserved Cambrian fossils from Chengjiang, China, show that even primitive vertebrates (fishlike forms that still lacked both the jaws and fins that would evolve later) were already present.

Canada's Burgess Shale and China's Chengjiang fossils paint vivid anatomical portraits of the spectacular Middle Cambrian soft-bodied animals that essentially document the completion of the Cambrian explosion. But the radiation of animals was not instantaneous; metaphors of explosions can mislead. Animal fossils cross the Precambrian-Cambrian boundary. I have scrabbled over the rocks of the White Inyo Mountains of California and seen the suddenly plentiful tracks of trilobites and worms, small shelly parts of otherwise soft-bodied animals, and inscrutable sponge-like reef builder forms called achaeocyathids, all living in profusion soon after the start of the Cambrian, 542 million years ago. Other signs go deeper. Below these Cambrian animals are the first tracks of living animals moving on a mud sea floor. The first fossil evidence of the most primitive living animals are now glimpsed from nearly 700-million-year-old chemical fossils of steroids produced only by sponges.

The oldest fossil record of animal embryos rivals that of adult body fossils and goes back about 575 million years in the Doushantuo Formation of China. These fossils, slightly less than a millimeter large, look identical to the cleavage-stage embryos of some modern animals. They somehow have been rapidly transformed from living cells to the mineral apatite. The creatures that produced these embryos were part of an enigmatic fossil fauna that dates from about 575 to 550 million years ago, just predating the first evidence of bilateral animals. These animals are known as the Ediacara fauna, named for the site in South Australia where their fossils were first discovered, but they occur worldwide. The appearance of animal hard parts in the fossil record is smeared over several million years. The tracks of the first bilaterian animals appear at the end of the Precambrian and became more elaborate in the ensuing Cambrian as these animals rapidly evolved new capabilities.

As for the predecessors of the Cambrian animals, I had first read about the Precambrian Ediacara fossils as an undergraduate in 1961 in a surprising cover article in *Scientific American* by the Australian paleontologist Martin Glaessner. He revealed the (to me at least) startling news of Precambrian animals, which he restored as a rich fauna of jellyfish, fronds, and worms, all related to living groups. That animals existed before the Cambrian was blockbuster stuff. I longed to see this fantastic fossil seabed that lay unimaginably remote from familiar Penn State, in the desert of south-central Australia. Glaessner's interpretations were later seen as naïve, and in reaction the jellies were replaced by suggestions that they were gigantic "quilted" protists, or undetermined thin creatures harboring algae, or perhaps tough lichen-like forms. Imagination was unleashed to flutter in nearly every direction other than animals. Recently, with more understanding of animal relationships and the fossils, there has been a return to a sober view more akin to Glaessner's. The Ediacarans were animals, but more primitive than he had conceived. Ediacarans are the remains of large and soft-bodied creatures. Their fossils have been hard for paleontologists to interpret, but they would appear most likely to be related to the living ctenophores and to cnidarians, which we know as jellyfish, hydras, anemones, and sea pens. This makes sense, as the cnidarians are more primitive in organization of their body plans than bilaterians.

WHY BOTHER WITH FOSSILIZED WORMS?

Most of the Earth's fossil record is made up of skeletons, teeth, and shells. These are the hard parts of animals and those most likely to be preserved as sediment turns into stone under the heat and pressure of deep burial. Sometimes these fossils litter quarries and eroding hillsides and attract scientific attention. This was the fossil record that Darwin was familiar with when he despaired of solving the dilemma of the apparently sudden appearance of large and complex Cambrian animals. No fossil record was visible in older rocks. Where Darwin fretted, creationists cavorted. The answer began to be teased out when a century after *Origin of Species,* fossils of cellular life, microbes, were discovered in microscopic sections of stromatolites, layered algal structures that have been fossilized. Stromatolites reach back over three billion years into the Precambrian and still live today in a few special places, like the Shark's Bay lagoon of Western Australia. When petrified, they are microbial gardens preserved in stone. Those discoveries showed that life long predated the Cambrian, but that life was still just microbes. The vital record of early animals was found at about the same time, but not in the form of bones, teeth, and shells. The first Precambrian animal fossils, those of the Ediacaran, had no skeletons. Body hard parts evolved just at the beginning of the Cambrian, and even after skeletons evolved, the majority of animal species had no hard parts. These animals were not all "worms." The earliest vertebrates, mollusks, and arthropods were all soft bodied, as were other major complex forms.

Although bones, teeth, and shells have documented most of the history of animal evolution for us, hard parts of complex creatures can only record a small fraction of the total anatomy of the animal. How true this is has been shown from the discoveries of fossil of elements of earliest Cambrian skeletons. Small skeletal fossils called "small shellies" were clearly parts of animals but could not be assigned to any specific kind of animal until complete specimens that included preserved soft body parts as well as skeletons were found. These early creatures were so alien that the hard parts had no correspondence to living animals, and so these skeletons could not identify the organism. Thus data from soft tissues contained the vital information. The more unusual the organism is to modern eyes, the more important the preservation of soft-bodied structure will be to deciding who their living relatives are. We can also gauge the importance of soft tissue by thinking about how much more we have learned from preserved soft parts about much better known large animals. Dinosaurs bring to mind big bones. Many more were small, and we now know from preserved tissue that many possessed feathers. We even have data on colors and color patterns – all soft-bodied information. Some kinds of soft-bodied preservation are unexpected and show unexpected, exquisite detail. The many extinct insects and spiders known only by being preserved in amber are gems within gems.

RIDDLES

There are three parts to the puzzle of animal origins. The first is the question of how animals evolved from single-celled ancestors. Because animals are made up of large numbers and many different kinds of cells, there is the problem of how to account for the origins of genes and developmental mechanisms that control the complexity arising in every generation as part of development from a single cell – the fertilized egg. How did development originate? The second question has to do with time and preservation. What if animals appeared much earlier than the Cambrian but were soft-bodied and small and thus failed to enter the fossil record, making them invisible to us for many millions of years of their history? Third is the question of an ecological why. The most extraordinary fact about the rise of animal life on Earth is not that it happened so fast, but that it took so long. The earliest fossils of bacteria date back to 3.5 billion years ago. The first fossil signs of animals don't appear until the planet is already 4 billion years old. Even if the first animals actually arose some millions of years before the radiation painted by their fossil record, there are still billions of years before that without any hint of their existence. Just what took so long? What happened to provide the conditions in which animal life could evolve? Why was Earth not hospitable as a home to animals sooner in its long history?

These questions have been my favorite puzzles for the past few years. The question of how development evolved in the Cambrian radiation blends into our earlier studies on how living embryos evolve. It had to have happened in the most primitive of multicellular animals. What were they? Sponges may be the oldest and simplest known animals, made up of the fewest kinds of cells. The idea that our earliest ancestors might have been sponges may not be appealing to everyone, but that's what genes suggest. Yet the question of who's on first is not yet completely resolved. Some investigators using molecular tools have suggested that sponges are likely the most primitive of living animals, and they have found that their closest relatives are single-celled creatures, protists called choanoflagellates, that look like the collar cells so characteristic of sponges. These modern singled-celled organisms presumed to resemble the ancestors of multicellular animals have been found to possess much of the molecular machinery that makes multicellular life possible. Bernie Degnan of the University of Queensland in Australia and his collaborators studying the sponge genome have shown that sponges contain an even larger suite of animal genes. So it would appear that much of the evolution of more complex animals involved the use of preexisting genes co-opted for new purposes. The remaining uncertainty about what the most basal animal lineage is arises because it's hard for the gene studies done so far to choose from among three primitive groups of animals, placozoans, ctenophores, and sponges, which deserves the crown for most primitive. Precambrian ancestors have been suggested for all three. This debate will make for some exciting future results before the animal ancestor finally fades to a one-line entry in a textbook.

All thirty-four animal phyla originated in the sea. Therefore, the marine phyla are particularly important for understanding how animals evolved, because they possess pretty much all the range of animal body plans seen in the Cambrian radiation. Only two phyla, arthropods and vertebrates, became success stories on land as large animals adapted to living in dry places. The arthropods on land are almost all insects and spiders. The insects predominate, with their millions of species, as yet mostly undescribed. There are a few other phyla on land represented by earthworms, snails, nematodes, and three more obscure phyla in damp places. But it's the sea that is home to nearly all the "minor" phyla, now with few species, like brachiopods, that were once dominant in the early history of animals. Most of the deep diversity of animal body plans and varieties of developmental processes still resides in the sea.

The sea runs riot in weird larvae and in the radical and fast evolution of new larval forms. With my students Belinda Sly and Meg Snoke, I proposed in 2003 that the earliest animals were all direct developers without distinct larvae, and that elaborate and distinct feeding larvae evolved later. Their evolution was prompted by changes in planktonic food sources and other factors in marine ecology during the Cambrian that favored feeding larvae. We tied the evolution of feeding, indirect-developing larvae to a process of co-option of genes used in the adult to take on new roles in development of the evolving larva. Co-option led to the construction of new gene regulatory networks that enabled the gain of swimming ciliary bands, guts, mouths, and other structures found in feeding larvae. The genes and the tissues they produced in adult development already existed, and they were available for new patterns of expression at an earlier time in development. Animal lineages that ancestrally developed directly were able to evolve an entirely new way of developing that let them exploit ecological opportunities not available to the adult. Our hypothesis is consistent with what we know about the evolutionary branching pattern of the phyla and what we know about fossil evidence for larvae from the Cambrian. In some phyla, the first distinct feeding larval body plans appear several millions of years after the basic body plans of animals had already diverged. Thus there was a second round of body plan evolution, that of larval body plans distinct from the adult body plans. This is another unexpected twist to the evolutionary history of animal origins.

Evidence from living animals and their genes and development gives us insights that can be compared to a rare fossil record. The fossils of early soft-bodied animals are extraordinarily special and valuable. Without complete soft body preservation, we would not even begin to guess at the existence of the extraordinary and enigmatic meter-long swimming animal *Anomalocaris,* which is a relative of the ancestors of both living peripatus and arthropods. *Anomalocaris* was thought to have been the great white shark of the Cambrian, but while I was writing this, James Hagadorn of the Denver Museum of Nature and Science presented a new interpretation of it as having a soft sucker mouth and unknown but more modest dining habits. Many more ancestral soft-bodied animals have been discovered, giving us a remarkable picture of the living Cambrian. The even earlier animals that arose in the late Precambrian are more problematic. The Ediacara fauna gives us one window into a few million years near the end of the Precambrian, during which a unique animal fauna thrived, but we know that millions of years of evolution had to have preceded them. We also know that much evolution of animal body plans had to have occurred between Ediacaran times and the classic Cambrian faunas. The past is always elusive, and its album still has lots of blank pages. We don't yet understand how soft-bodied fossils were preserved.

BECOMING A FOSSIL

We gaze at massive dinosaur skeletons. Yet much is lacking in them. A skeleton preserves only a fraction of the total structure of the living animal. Its soft tissues, metabolism, behavior, development, and genes made up the bulk of information in its body, and they have vanished. Only in a few special cases does the record reveal more. For poorly understood reasons, there are conditions that have allowed preservation of soft tissues at some lucky moments in geological time. These "soft-bodied" fossils contain an extraordinary amount of information. By chance, many animals of the Cambrian radiation are preserved in this way, and in some cases, even their embryos and larvae have been preserved. Because they are so wonderfully rare, geologists have given a special name to sites where well-preserved soft-bodied fossils appear: lagerstätten, literally, a "resting place."

In order to understand soft-bodied fossils, we have to have some idea of what kinds of organisms they represented and we have to be able to reconstruct how they formed. Some are not simply flattened impressions as once thought but record three-dimensional information. Soft-bodied fossils are usually the mineralized replacements of tissue. That's a step removed from the soft tissue anatomy of the living organism. Understanding the translation from living tissue to a mineralized fossil will help us understand what was once there and, equally important, understand what cannot be preserved. The biases of fossilization processes color what we think was present. We will never have a full record, so if there are systematic losses of information, we want to know. Mineralization requires certain chemical and, now we have found, microbiological conditions. We are only beginning to understand how the processes involved work and what biases might distort what kinds of organisms and structures we find preserved. Fossilization thus requires understanding the organismal biases, the chemistry, and the biology of the way microbes produce a fossil. And we need to understand how the processes can work so rapidly, because small, soft dead things just don't hang around very long on the sea floor. When the special conditions are present, these soft remains are preserved rather than decaying, the fate of the vast majority of creatures when they die. These fossils are the jewels of the fossil record, the rarest of the rare.

FOSSILIZING SOAP BUBBLES

Most mineralized fossils are tiny, in the range of millimeters in size or thickness. Some of these are exquisite little animals with legs and mouthparts. Some are developmental stages of late Cambrian arthropods. These creatures are preserved by replacement of tissue by the mineral apatite, a form of calcium phosphate that makes crystals smaller than a bacterium. The tiny crystals are in effect pixels that determine the fine detail that can be preserved, and structures the size of animal cells should in theory be preserved in this way. The discovery of fossils of what seem to be the earliest animal embryos from the late Precambrian and the early Cambrian opened a whole new window into understanding how animal development originated. Could we find modern analogs that would let us answer questions about how early animals developed and how they were fossilized? Tiny Cambrian larvae and small animals are known as highly detailed mineralized soft tissue fossils. A decade ago, fossils preserved as apatite of what appear to be stages of cleavage of animal embryos were discovered from the Ediacarian age rocks at Doushantuo, China. But were the Precambrian embryos from Doushantuo really embryos? Maybe they were fossils of something else of about the same size, such as giant sulfur bacteria or algae.

Beth and I took on the study of how the embryos might have been preserved in 2005 as a result of a challenge by our British collaborator, paleontologist Phil Donoghue, who asked us if we could think of a way our *H. erythrogramma* embryos might become fossilizable. They are about the same size as fossil Precambrian and Cambrian embryos and larvae and would seem to be a good model. At first sight, we thought it wasn't feasible. If you kill a marine embryo and leave it in seawater, it will begin to degrade within hours. How could it be stabilized to undergo a process that would take at least weeks? It seemed like a problem in fossilizing soap bubbles.

We took the challenge as being one of cellular biochemistry. The degradation of dead embryos is fast, and so it would seem to involve enzymes from within the embryo itself. These destructive enzymes would only be released from the cellular compartments that held them, or activated once the embryo cells died. This process of self-destruction is called "autolysis," and we thought that preventing it would be the key to passing any soft-bodied organism into a fossilization process. Surprisingly, we were correct, and the problem was easily solved. We found that we could prevent autolysis by treating killed embryos with a reagent known to unfold enzymes so they lost their correct molecular shapes and could not act. Such embryos can be stored for months without any autolysis. We also discovered that two kinds of environments present in the ancient oceans would also block autolysis: oxygen-free environments as would exist if an embryo sank into anaerobic water or environments with high concentrations of hydrogen sulfide. That gas was present in high amounts in anoxic basins in the late Precambrian and early Cambrian. It is still formed in some environments by certain microbes, and it's even pumped out of hydrothermal vents at sites on the sea floor. So reproducing in the lab the special conditions for the first step to fossilization in a late Precambrian ocean is feasible.

What then? Bacteria are pretty generally cast as insatiable munchers that should cause rapid destruction of a dead soft-bodied animal or embryo – and mostly they do just that. We looked in the simplest of ways to see what havoc the bacteria wrought and found that the presumed villains were actually the heroes. We exposed dead embryos, which we had treated to prevent self-destruction, to natural seawater. Seawater contains lots of bacteria. We thought that bacteria would quickly destroy embryos and that we would have to find environmental conditions that would block bacterial growth. We then examined the embryos that we had exposed to bacteria by scanning electron microscopy, which would show us where the bacteria were and how much degradation they had caused. A good hypothesis is one that's easy to prove wrong. We had based our thinking on a such a "good hypothesis," and it was mistaken.

Serendipity had struck. Some seawater bacteria quickly ate the embryos, but in an unusual way. They replaced the detailed structures of the embryos and their cells with a replica that retained the shape of the embryo. The embryo replica was stable, but no embryo was left. Instead, only the form of the embryo remained, composed entirely of a biofilm made of bacteria and the "glue" they secrete. We called this mode of shape preservation "pseudomorphing" because the delicate form of an organic structure is kept but is translated into something else far more stable. Not all bacteria make pseudomorphs, but we discovered some strains of marine bacteria that do it well.

The next question to ask is, if soft tissues can be readily replaced by the bacterial mold of the tissue, can the bacteria that make up the pseudomorph somehow carry out the next step? Can they cause the deposition of minerals in a short time? If so, can we get the right minerals that would be required for a well-preserved natural fossil to form? By controlling the chemistry and microbes in which we suspend dead embryos, we have found that we can control whether minerals form and what kind of mineral is deposited. We thus have been able to get rapid deposition of the most important mineral, calcium phosphate, although we have not yet been able to replace the embryo by a fully mineralized replica. Paleontologists have noted that the initial critical steps of fossilization of soft-bodied tissues must happen very quickly. Our experiments bear this idea out. The processes that lead to stabilization and initial mineralization take only days under the right conditions. I imagine that those who think that the fossil record comes from the great flood will wishfully take heart from hearing that fossilization processes can take place in less than forty days and forty nights.

We are now entering into genomic studies of bacterial strains that take our dead embryos through the steps leading to fossilization. We aim to find out what genes are involved in the microbial processes that carry out the individual steps. Modern genomics allows us to completely sequence the genes of the microbes we find to be important and to then investigate the role of specific genes in the bacterial mediated steps of fossilization. Not all related species of bacteria carry out all steps equally well, which allows for gene screening. Once genes have been identified, they can be individually inactivated to directly look at their roles. The marine bacteria that we find to be heavily implicated in fossilization steps belong to a group known as the gamma-proteobacteria. A number of these bacteria have close associations with living marine animal embryos and development, and in some cases they can even be required for development. Some strains even trigger larval metamorphosis. Perhaps the most famous is the symbiotic strain of *Vibrio* that provides the light that shines from squid light organs. Analogous symbioses lie in the association of lobster eggs with a related bacterium that covers the eggs. In the absence of these bacteria, fungi kill the eggs. In the case of killed embryos, the bacteria already present in a biofilm on the surface have first dibs on invading and consuming the embryos. We suspect that we will be able to learn about what must be ancient interactions between embryos and microbes, probably dating to the late Precambrian origins of animal embryos.

INTO THE GARDEN OF EDIACARA

In 2007, a flight on Qantas from Sydney to South Australia would fulfill my Ediacara dreams of forty-plus years. Beth and I had a scientific reason to go as well. We wanted to better understand the paleontological background for comprehending soft-bodied fossilization. We also had to put the earliest fossil embryos into context as they provide us the earliest direct evidence about how the first animals carried out development. Much of the study of the Ediacara fossils has taken place in the South Australia Museum in Adelaide. This gray stone castle that stands under blue Australian skies is the gateway to entering the lost world of Ediacara. We flew into Adelaide to visit with Jim Gehling, a paleontologist who has devoted his career to these fossils. The museum has an up-to-date exhibit of stunning Ediacara fossils and stimulating restorations of the imagined living beings. Ana Glavinic, who studied how opalized fossils form, and Jo Bain, who prepared the spectacular exhibit, showed us around.

The Ediacaran animals are preserved as sharp impressions in a fine-grained hard quartzite. Most Cambrian fossils are relatively small, but the striking thing about seeing their fossils is to have the realization of how large some of the Ediacarans were. A meter-long specimen of the ribbed discoidal fossil *Dickinsonia* was bolted to the wall – a huge but incredibly flat creature unlike any animal today. Next there was a slab with a pair of nearly two-foot-long fronds of a creature called *Charniodiscus*. These are so well preserved that it is possible to see what look like individual polyps on the fronds, which along with the frond and holdfast structure implies that they are cousins to living sea pens. This massive specimen had become notorious when, a few years earlier, it was stolen from a cliff in the Flinders National Park. It seems odd to leave a national scientific treasure unguarded, sitting outside to flake in the weather or be stolen. This one was filched by a hardy band of fossil pirates who included a former museum paleontology technician. The heist was no small job. The thieves used a portable diamond saw to remove the slab from the immense block of hard quartzite of which it was a part. The heavy but fragile fossil was then rappelled down the cliff and smuggled out of Australia. Nine years later the museum got a tip that the missing booty was for sale in Tokyo for a reported $750,000. The Australian government filed a complaint with Japan and the fossil was returned unharmed and professionally cleaned. That's when the impressions of structures that look like polyps were discovered on the fronds.

The interpretations of Ediacara animals remain uncertain in some ways but are reminiscent of later animals in others. We can see from impression of wrinkled or folded-over specimens of *Dickinsonia* that the creature was flexible and pancake flat. There are marks on slabs that preserve the algal layer that lay thick on the sea floor, suggesting that *Dickinsonia* may have moved slowly over the microbial mats absorbing nutrients. One recent suggestion by Erik Sperling and Jakob Vinther is that it was a gigantic version of a simple and tiny still living organism called a placozoan. These simple creatures seem to lie in phylogeny between sponges and more complex animals. Beth and I in 1970 published a theoretical paper in *Nature* on how the Ediacara animals, which were flat creatures and likely were primitive animals, could have respired by diffusion at their surfaces. This marked the first of our occasional joint projects over the next forty-plus years on the nature of the first animals and how they were preserved as fossils. I think that first paper still catches the essence of how these animals functioned both for gas exchange and absorption of nutrients.

As for a record of when the first bilateral animals arose, Jim Gehling set in front of us Ediacara fossils that exactly resembled tiny trilobites. He suggested that they were early forms of segmented animals, possibly arthropods. It was hard to be certain. Another fossil, *Kimberella*, seemed to be a bilateral creature that fed like a mollusk by scraping algae off the sea floor. Perhaps the Ediacaran animals comprised a jelly world of primitive soft-bodied animals of several basal evolutionary lineages of animals, perhaps even including extinct ones that left no descendants. They dwelt in a world before predation, eyes, and the hard body armor that protected the hunted from the hunter arrived on the scene – in what Mark McMenamin once called the "Garden of Ediacara" for its seeming ecological innocence.

We were particularly curious as to how these strange flat animals had been preserved. Jim showed us the tops and bottoms of slabs of the Ediacara quartzite. The top was clean, pale quartzite that bore ripple marks like modern sand. These surfaces recorded a long-ago bare white-sand sea floor gently rippled by currents. The lower surfaces of the same slabs were rust colored, indicating the presence of iron. Ediacara fossils are found on the slab bottoms where the living animals were suddenly buried when they were covered and smothered by sand dropped by storm surges. They ended up entombed under a thick layer of sand that became the new sea floor. The undersides record the event. The slab bottoms bear two kinds of fossils, the negative impressions of animals and thick microbial mats. The impressions left by the microbial mats are called "elephant skin." Jim has noted that the rust is the trace of oxidized iron sulfide, pyrite, which formed death masks that preserved the impressions. Decay of the microbial mats provided the anaerobic chemical environment needed to produce rapid iron sulfide mineralization. Robert Hooke in the 1600s was the first to use the microscope to examine fossils, and he proposed that fossils came into being by replacement of animal or plant tissues by minerals. The study of just how the dead become fossils now has a name, taphonomy.

The Ediacara fossil bed lies in the Flinders Range. We drove north from Adelaide through the wine country and into the ever-drier landscape of interior Australia. The road passed through a zone of racing pink dust devils approaching what turned out to be a boutique mountain range with a perfection of scale and starkly beautiful scenery. We stayed at a small resort tucked away at the north side of the base of a stone monolith called Wilpena Pound. The pound is an immense shallow stone bowl that glows red at sunset. Its sides are made of a succession of cliffs, each representing a layer of hard Precambrian quartzite. The line of Flinders Range peaks runs north from Wilpena Pound like the tail of an immense rocky tadpole. At first glance people think the pound is a volcanic crater, but it's not. It is a geological feature called a syncline and is the remnant of once-flat rock layers bent to form what was the bottom of a valley in a great mountain chain. The former ancient mountain peaks have been ground down by time, leaving today's remnant Flinders Range.

Our drive extended into the dimness of twilight. We noticed that kangaroos were gathering silently by the roadside, still and partially concealed by brush. They would sit calmly, watching us approach with their big, wide empty eyes. At exactly the right moment, one or another would inexplicably throw itself into the path of the car. The roadside was littered with their recently deceased predecessors. Kamikaze kangaroos. Hitting a kangaroo in midair and having it crash through the windshield is not recommended. We decided to drive more slowly through the shadowy kangaroo roadway. In daylight, while the surviving kangaroos slept, we would discover that the roos were replaced in their death-defying efforts by the even larger flightless birds, emus, which liked to dash out in front of us and race the car down the center of the road.

The Flinders Range points north into the red Simpson Desert of central Australia. The mountains catch enough moisture to be well vegetated, although the plains on either side run to extreme desiccation, with white desert salt pans off to the east and west. The gorges that cut through the ranges spread out a geological map that lays out a detailed rock record of the late Precambrian world, the end of the world of microbial slime through to the dramatic appearance of animals. The best way to visit the age of the first animals is to drive down the four-wheel track through Brachina Gorge from east to west. There you first pass through 800-million-year-old Late Precambrian rock layers that preserve fossil stromatolites the size of giant tortoises. Then one moves up in time until a decisive break, the traces of the last glaciation event of a series of late

Precambrian glaciations known as "snowball Earth." The great snowball Earth glaciers apparently extended nearly to the equator.

The east entrance to Brachina Gorge cuts through a rock layer called the Elatina formation, which was deposited by the last snowball Earth glaciation. It is composed of unsorted materials where glacially rounded stones of different sizes are dumped unceremoniously together with what once was mud – a solidified rock-and-silt pudding called "tillite." Glaciations were ended suddenly about 635 million years ago by a big global warming event that left its traces right above the Elatina glacial rocks. This upper formation is dubbed the Nuccaleena Formation. It is an unusually dense limestone called a "cap carbonate," which preserves rapid crystal growth and shows that a high level of CO_2 had been released into the atmosphere. This event warmed the atmosphere, turning the frozen snowball Earth into a hothouse world. Accelerated erosion of rocks by CO_2 made calcium available and massive amounts of limestone were deposited in the sea. These two modest-looking rock layers mark a stupendous change in the biosphere. Similar rock layers in Namibia, Australia, and China show the worldwide span of these peculiar glacial events and their end.

From about 1.8 billion years ago to near the end of the snowball Earth era, the oceans remained their ancient anaerobic condition and had a chemistry controlled by hydrogen sulfide. This peculiar marine condition is called the "Canfield Ocean" in honor of geochemist Donald Canfield, who first proposed it in 1998. By 700 million years ago, near the time the episodes of snowball Earth were drawing to a close, the grip of anaerobic seas was starting to break and high levels of oxygen for the first time in four billion years gave the earth an atmosphere similar to ours. The oceans at the end of the Proterozoic became heterogeneous, with an oxygenated layer overlying sulfidic waters. Oxygen-rich seas opened an astonishing revolution in the Earth's chemistry and made animal life possible. Animals, which require high oxygen levels, began their rapid evolution.

The trip west into Brachina Gorge follows the flow of water downstream but is a geological ascent through a winding tight canyon of tilted colorful rocks. The famous pinkish Rawnsley Quartzite, which bears the Ediacara fossils, lies midway between the flowering of microbes in Precambrian stromatolites and the blossoming of animals in the Cambrian. A few Ediacara fossils are visible on rich brown undersurfaces of the rock layers in the cliff of the classic site. Structures called "sole marks" and "slump rolls" on the undersurface of thick quartzite layers tell of fast burial of the soft sea floor by large storm events. These floods of suspended sand buried the sea floor life instantly in place, preserving it on the undersides of the storm beds. The pale, rippled top surfaces preserve the new sea floors formed after storm events. These rocks are about 550 million years old. Rocks 15 to 20 million years younger lie farther along downstream. They are important because they show exuberant Cambrian animal life. These are familiar fossils, the large intricate skeletons of the sponge-like archaeocyathids that built reefs. They are strikingly visible as black cross sections in the hard, off white Early Cambrian rocks polished smooth by running stream water.

Wildlife abounds in the gorge, and as always in Australia, it thrives in the form of oddly appealing characters. The largest animals are the man-sized emus. They are silent but common, and they supply a good value in dinosaur accompaniment for a geology field trip. Male emus herd their numerous downy chick charges, while females, who consider their duty done once they have laid the eggs, roam responsibility free. There are abundant kangaroos, but not the familiar eastern grays of New South Wales. The ones here are a longhaired species known as "euros." They lie passive in the midday heat or, later in the afternoon, gather around one of the creek water holes for a drink. As I scanned the sunny Ediacaran fossil cliff face, I had the good luck to see a yellow-footed rock wallaby move nimbly across the cliff tops. This is an extremely rare species. He was a gorgeous animal, glowing in the light, a sort of elegantly patterned small kangaroo, brown, black, white, and yellow on its upper body, with yellow legs and a tail banded in yellow and black. He paused to look down at me and disappeared leisurely, lightly bounding over the sun-baked scree. I was left gazing up at the empty rock face.

Noise in the Flinders is generously supplied in the form of shrieks by gangs of sulfur-crested cockatoos and flocks of little corellas. Downright oddness comes in unexpected surprises – such as a strange young bird, a frogmouth, patiently waiting by the side of the track for its mother to stop by with a lizard or mouse treat. Lizards were represented by the placid and heavily scaled giant shingleback skinks, warming themselves in the morning light. We saw none of the most mammal-like of lizards, the big carnivorous goannas – a pity. Discomfort was plentifully available courtesy of the ubiquitous flies so importuning of body fluids in the heat. And then there were the plants – starting with the biggest and most obvious, the monstrous baobab-like river gums that line the watercourses in the Flinders. They are distinctive but seem to be the ecological equivalents of our western cottonwoods. But once again, Australia surprises with a convergence of some of its plants with the familiar forms of the Northern Hemisphere. The hillsides in places looked like Wyoming, with a cover of sage bushes and scattered pines. Only, they weren't. The "sage" was an amaranth the color of sage. The "pines" that cover the Flinders slopes are a relative of cypress called *Callitris*. Australia has another, better-known and widespread pine mimic, this time actually a flowering plant, not a conifer: *Casuarina,* the "she-oak," which look just like a long-needled pine. *Casuarina* "needles" are actually slender, green, jointed branches that serve as leaves. There is nothing oaklike about these trees at all, so why "she-oak"? Because homesick settlers used the timber in woodworking and seem to have found it not as hard as the familiar English oak.

Wildlife is celebrated in a culinary way on the dry plains just west and north of the mountains. Just off the highway leading north to the Ajax Mine, where Cambrian fossils occur, lies a famous haven for thirsty geologists wandering the dry desert, the extraordinary Prairie Hotel. There in the cool sunny dining room, along with beer and wine, and work by aboriginal painters, is served their luncheon special, a kangaroo cold cut, camel sausage, and emu paté platter. It is zoologically diverse and tasty, and blessedly a far cry from the tough schnitzel that makes up most of the offerings of the hotels in the one-dog towns out this way. The dromedary camels that supply the nice sausages were introduced to Australia in the nineteenth century, and they have made the plains their own. Just one more introduced species that seemed like a good idea at the time.

The Australian desert had one more reward for us. In the Flinders Range, as in the deep red center near Ayres Rock, the Milky Way stretches brilliant in the light of its thousands of visible stars spanning from horizon to horizon. Watching it, we felt as though it could draw us bodily up into the galactic center and the universe beyond. The Greater and Lesser Magellanic Clouds, two relatively close satellite galaxies of the Milky Way, showed as fuzzy patches of light. Galaxies, giant discs of stars like the Milky Way, are a class of celestial objects simply not visible in the Northern Hemisphere. Even the Milky Way is becoming harder to see. So much of our sky in the United States is made opaque by city lights that, beyond the Moon and a few bright planets and stars, we hardly see the sky at all.

PART THREE

Strange New World

AMERICA'S REJECTION OF EVOLUTION ON RELIGIOUS GROUNDS has its own peculiar flavor and history. I've watched with dismay as evolution denial has metamorphosed into a general dismissal of scientific reasoning and, worse, turned virulent as conservative religion and conservative politics have united. Yet comprehension of evolution has become ever more essential to understanding nature, and necessary if we are to have any prospect of preserving our natural world.

Ignorance is the soil in which belief in miracles grows.

Robert Green Ingersoll

EIGHTEEN

Darwin's Day in Court

EVOLUTION MEETS RELIGION IN AMERICA

When Galileo turned his telescope to the sky in 1609, he revolutionized our thinking about our place in the universe as surely as in modern times Darwin's natural selection would give us a new view of our biological origins in nature. Galileo's startling discovery was that Copernicus had it right. The Earth was not the center of the solar system, nor was a fixed Earth the center of the universe. In 1632 Galileo published a defense of his work and was tried for heresy and condemned by the Catholic Church to spend the remainder of his life under house arrest. In 1992, Pope John Paul II acknowledged the church's errors in the matter. In 2009, the Pontifical Academy of Sciences concluded, "The extraordinary progress in our understanding of evolution and the place of man in nature should be shared with everyone. . . . Furthermore, scientists have a clear responsibility to contribute to the quality of education, especially as regards the subject of evolution." The validity of science does not depend on religious approval, but amicable understanding doesn't seem like a bad idea. That's an unfortunately iffy proposition for science co-existing with at least some of religion in America.

German general Erich Ludendorff referred to the poorly commanded British soldiers in World War I's Battle of the Somme as "lions led by donkeys." We can co-opt this observation for reference to Americans led by the recent George W. Bush administration. During those eight years we watched religious right fundamentalist Christian dogma override the best scientific judgment available to federal advisory panels and agencies. The free application of dogma grievously affected making rational policy decisions on sex education, birth control, world population policy, climate change, abortion, and stem cell research – not to mention attempts, which continue apace, to promote teaching creationism in schools. The Bush administration was not interested in effective policy, only in effective politics. To repay the administration's so-called base, religion of a particular fundamentalist Christian stripe was brought into government. Television con-men evangelists affected American science policy through their large audiences and their growing influence in the Republican Party. Their efforts are continuing, and we reached the absurd point where the most prominent Republican candidates for the 2012 presidential election proudly trumpeted their piety and their disdain for the sciences of evolution and global warming.

Most of America's early religious roots lie in Protestant denominations such as Episcopalians, Methodists, and Baptists in the eighteenth century, to which were added Catholics in the nineteenth and twentieth centuries. Atheists and agnostics also account for a substantial proportion of Americans; polls show that over 15 percent of people identify themselves this way. Religious diversity prevails. Many offshoots of Christianity arose in America and are exceedingly varied. Pentecostals, Assemblies of God, Primitive Baptists, Southern Baptists, Nazarenes, Mormons, Jehovah's Witnesses, Seventh-day Adventists, Christian Scientists, and Appalachian snake-handling cults have exploded into an astonishing spectrum of religious practice, with some notable accompanying zaniness. It is the culmination of a century and a half of American theological creativity. Many of these sects take a fundamentalist view of Scripture, and firm belief in creationist theology is prevalent.

Creationism is not a scientific concept. It is a purely religious idea, and it originated among believers of the literal truth of the Bible. Creationism was a reaction to the challenge evolution posed to religious ideas as to the cherished special place of humanity. Modern Christian fundamentalism arose as a largely American phenomenon, born in the early twentieth century with the publication of a series of booklets containing the "fundamentals" of the faith. These fundamentals were composed to combat liberal biblical scholarship, which revealed the human authorship of and inconsistencies in the most basic texts and teachings of this supposedly infallible book. Fundamentalist sects are intolerant and antidemocratic by the nature of their belief in an absolute truth and in the authoritarianism that arises from having to adhere to the word of God as spoken from the pulpit by those divinely favored souls who expound it, the preachers. The writers of the fundamentals demanded that every word of the Bible had to be taken literally or faith would lose its meaning. A public education system that taught that humans descended from other animal species was going to be a noticeable anathema, and conflict became inevitable. Because the birth of fundamentalisms occurred in the ever-litigious United States, much of this century-long battle has been fought out in the courts.

However, creationist religious beliefs themselves are not sufficient to account for the current problem creationism now poses for education and public understanding of science. The influence of creationism is reinforced by a growing religious movement, dominionism, which abhors our secular form of government. Dominionism means to reform society so that Christians have total charge over government and control of public life, a delightfully self-serving proposition. In the view of dominionists, separation of church and state is merely a falsehood concocted and spread by secularists and was not actually a part of the U.S. Constitution, despite the incontrovertible wording of the establishment clause of the First Amendment. In the visionary and revisionist account of American history created by the dominionists, the founders intended the country to be an explicitly Christian nation. This is a mischievous enough idea by itself, but it represents only the milder end of the spectrum of dominionism. At the most extreme end there is Christian reconstructionism. Here lives the frighteningly virulent idea that "biblical law" must be instituted to govern America. The proposition is no longer merely one as extreme as Christian control of society and government. What reconstructionists propose is an entire legal system based on the rules laid down in the Bible, which they believe should be substituted entirely for the Constitution and our current code of law.

On that bright day, we will share the glories of living under the bloodthirsty laws of Deuteronomy and Leviticus, including the execution of disrespectful children, minding in our dress the stricture against wearing a garment containing both wool and linen, enjoying the right to enslave heathens, and gathering for public events where we'll have a rousing fine time stoning sinners to death. Reconstructionists not only hold that these Bible-based Bronze Age laws should govern both American politics and private life but also campaign to make it so. The dominionists who fall anywhere in the spectrum of religious dominance are in fact the people the writers of the Constitution warned us about. Dominionist sentiment is prevalent among American fundamentalist Protestants, and the drive of dominionists for political domination is determined and profoundly antidemocratic and antirationalistic. Their views and activities as re-writers of American history are an attempt to control the present by rewriting the past. The ultra-Christian Founding Fathers so breathlessly hailed by dominionists didn't exist. Actual major figures in the founding of the American Republic, including Thomas Paine, James Madison, and Thomas Jefferson, knew the costs of religious strife from looking back to the history of the Hundred Years' War and the religious persecutions of minority sects in the American colonies. They made every effort to separate religion from government for the protection of both. Jefferson was concerned about religious interference in democracy. Here is Jefferson writing to the great explorer-scientist Alexander von Humboldt in 1813: "History, I believe, furnishes no example of a priest-ridden people maintaining a free civil government. This marks the lowest grade of ignorance, of which their civil as well as religious leaders will always avail themselves for their own purposes."

Political candidates and government officials have taken to pandering openly to religious conservatives and dominionists. Think of George H. W. Bush saying of nonbelieving Americans while running successfully for president in 1987, "I don't know that atheists should be considered as citizens, nor should they be considered patriots. This is one nation under God." I assume that he knew better but was willing to play to the bigots for political gain. Views like these reflect the exclusionist outlook of petty tribal religions. Think of the undoubtedly sincere but stunningly stupid comment of an American general, William Boykin, the deputy undersecretary of defense for intelligence in 2003, comparing his God to Allah: "I knew that my God was bigger than his. I knew that my God was a real God and his was an idol." He apparently did not realize that the word "Allah" refers to the same monotheistic God that he believes in.

One dominionist movement is the so-called New Apostolic Reformation, which is subscribed to by Sarah Palin, in 2008 the Republican candidate for the office "just a heartbeat from the presidency." In preparation for Jesus' return, this movement applies, among other innovations, the driving out of demons and witches and "spiritual warfare" aimed at seizing control of politics and all other societal institutions. The believers are fully confident that their prophets and apostles are fit to run the nation as deputies for God. Loony ideas indeed, but not so harmless when harnessed to politics. One of the originators of what is now this movement, William Branham, once gave a foundational sermon on the subject of the "serpent seed." This is the idea that Eve's children through Adam were upright citizens, but her child from sex with the serpent was the notorious Cain. According to Branham, Cain's descendents were "smart educated, intelligent people. . . . They were builders, inventors, scientists." This notion on the part of someone who wants to see his movement assume control of the country is not likely to comfort rationalists. Other dominionist groups operate on a higher socioeconomic level, quietly co-opting congressmen and senators to subvert democracy through developing a loyalty to something supposedly higher. An embarrassing series of revelations about the secretive C Street house have led to its being dubbed Washington's frat house for Jesus. We now have a glimpse into that quiet effort to create a Christian state from the top down.

These ideas are not held by a majority of religious people, who accept scientific ideas without damage to their religious beliefs. However, an open-eyed look at the extremists is important, because although they are, at least for now, a minority, they have developed effective propaganda outlets in megachurches and religious television networks and formed powerful political alliances. They influence the views of the wider society, and they thus impact our freedom to think rationally, to employ science effectively in public policy, and to teach science honestly. To see how we got to this moment in American history, when dominionist politicians reject science and openly call for the destruction of our constitutional right to separation of church and state, we need to start at a relatively innocent time in the process that has brought us to this point.

NO EVOLUTION, PLEASE – WE'RE TENNESSEANS

Darwin famously first entered the courtroom in Tennessee, during the summer of 1925. The year before, the state legislature had passed without debate a law, the Butler Act, which made it illegal in any state-funded school in Tennessee "to teach any theory that denies the story of the Divine Creation of man as taught in the Bible, and to teach instead that man has descended from a lower order of animals." A couple of local entrepreneurs looking to promote business in the sleepy town of Dayton conceived of the publicity stunt of mounting a court test of the Butler Act. They talked a young science teacher, John Scopes, into acting as the defendant. Scopes was young, unmarried, and game to take part, although from his own account, it's not clear just how much evolution he taught – if any. Scopes was not the school's official biology teacher; he only was a substitute teacher in biology while the regular teacher was out sick. But Scopes provided the opportunity to bring the test case. Science and religion could now duke it out in public view under the flying colors of small-town boosterism.

No matter the original motivation, the game of challenging the constitutionality of the Butler Act was taken up under the auspices of the American Civil Liberties Union. The case attracted newsworthy talent to the prosecution and to the defense. Creationist and former presidential candidate William Jennings Bryan joined the prosecution, and the famous, or infamous (depending on one's point of view), New York attorney Clarence Darrow joined the defense. Scientists volunteered to testify about the validity of evolution. However, that was not to be. Presiding Judge John Raulston seems to have had remarkably little imagination and declared scientific testimony to be irrelevant. This seemed like a blow at the time to the defense, but Raulston's ruling actually had the unanticipated bonus of resulting in the legendary face off between Darrow and Bryan in which Bryan allowed himself to be questioned as a witness by Darrow on the literal truth of the Bible. This battle of titans led to some remarkably funny testimony, with Bryan revealing his poor knowledge of science and his own doubts of the length of the creation day. The testimony produced such classics as Bryan's waffling on the issue of how the biblical age of the Earth was calculated:

> DARROW "What do you think?"
> BRYAN "I do not think about things I don't think about."
> DARROW "Do you think about things you do think about?"
> BRYAN "Sometimes . . ."

After the trial, Scopes refused to cash in on his day of fame. He received some financial help from supporters that allowed him to earn a master's degree in geology at the University of Chicago and later become an oil company geologist. Scopes late in life wrote an autobiography called *Center of the Storm,* a poorly known but revealing book. Few people seem to consider Scopes as more than a convenient pawn, a cipher. He was self-deprecating and says of himself that he was a passive observer in the trial. Nonetheless, he strikes me more as a private person who willingly took on his role in the legal challenge "Monkey Trial" as a matter of informed principle. His family background seems to have been a significant motivation. Scopes was the son of a self-taught labor leader, who turned up for the trial to show his support. Father and son were both agnostics. Scopes was a rationalist and made it clear that he had read *The Origin of Species.* He says that he would have taught evolution if the course had been his. Despite holding religious and social views that would have been regarded as radical in this fundamentalist county, he was well liked and attended church regularly during his time as a teacher in Dayton.

Scopes as defendant had to sit as a passive observer of the trial. However, he took an active role in the preparations for his defense. The debate as to who should lead the defense was ended by Scopes's choice of Clarence Darrow to be his trial attorney. Scopes well knew that Darrow was openly agnostic and that his reputation was as a controversial but brilliant lawyer. Scopes thought that he had the punch to take on William Jennings Bryan, who was a powerful speaker and operating on overwhelmingly friendly ground. The defense was not looking for an acquittal. They not only wanted to make a public case for evolution as valid science, but also wanted Scopes convicted. And so he duly was. This was the peculiarly happy outcome that would let Darrow take the case to the

Supreme Court. Alas, that was not to be. The brilliant verbal fireworks that cheered northern liberals and lit up the summer sky of 1925 were just harmless sheet lightning. A state court overturned Scopes's conviction on a technicality over how the fine was assessed. There would be no test of the substance of the case to advance to a federal court, nor would there be such a test anywhere for quite a long time. The Butler Act stood proudly for another forty years, and evolution slipped unnoticed from American high school science textbooks. However, the dramatization of the case in the 1955 play *Inherit the Wind,* and movie versions of it, have kept the vision of the trial alive in American memory.

Scopes wrote his personal account forty years after his gaudy show trial. The passage of time and subsequent knowledge of the history of the events may well have altered his memories of his role, but his recollections ring true, and I doubt that time altered his telling of his father's views, which so clearly had a strong influence on his own development. Scopes admirably did not use his book to disparage Bryan or Judge Raulston. He respected Bryan's passion for people. In fact, Scopes records an extraordinary incident that took place during a hot afternoon break in the trial. He, the defendant in a criminal case, went off to a swimming hole in the hills near Dayton for a swim with two prosecution lawyers, one of them Bryan's son. I was struck by the story and asked my trial lawyer son if this was standard procedure. He had a good laugh and told me he couldn't imagine such a thing happening now. It stands as an unusual human moment. Scopes, by the way, returned to Dayton in 1960, when he was invited to help promote the movie version of *Inherit the Wind.* Not so much had changed. Teachers still had to sign a pledge that they would not teach evolution. Scopes was denounced in a sermon by a local preacher, and the town soda shop offered a "Scopes Soda" for fifteen cents.

Bryan had been U.S. secretary of state and a presidential candidate. His political career is one that would appear almost impossible today, given the current link between right-wing politics and fundamentalist Christian sects. Bryan was liberal in his politics but conservative in his religion. His opposition to evolution was based on his belief that the acceptance of Darwinism had fueled German militarists in starting World War I. Thus his was a moral crusade, apparently based on the delusion that there was no evil in the world before 1859. I don't doubt Bryan's sincerity, but I wonder if he ever actually read the Old Testament in any critical way. Bryan died dramatically a few days after the trial, and a small fundamentalist college, William Jennings Bryan College, was established in his memory in Dayton, Tennessee. I remember once when I was an undergraduate looking at the college catalogue to see what went on there. My best finding was an activity called the Missionary Aviation Club. Makes you wonder. Still, Bryan's moral concerns can't be waved off. Ethics matter, but it is a question where they arise. Many religious people think ethics can only originate from the rules laid down in holy books, but that's false. Consider how the words attributed to God in the Bible reflect those of an evolving society. The Old Testament God demanded blood and burnt offerings and required stoning for such offenses as wearing a garment made of two kinds of thread. By the New Testament, Jesus speaks in a much more kindly and appealing way. Ethics, like a sense of religion itself, appears to be a behavioral characteristic evolved as part of the complex of traits of an intelligent social species.

A final note: In April, 2012, a full 87 years after the Scopes Trial, Tennessee passed a law that encourages teaching the "controversies" about "biological evolution, the chemical origins of life, global warming and human cloning," the usual targets of the religious right. The law will encourage the teaching of anti-evolution concepts drawn from creationism and intelligent design under the guise of non-religious critical questions. The motive is religious, and the result, intended or not, will be the undermining of science education. The law will inevitably be tested in court, and the trial just as inevitably will be called Scopes 2.

We are taking the dinosaurs back from the evolutionists.

Ken Ham, Creation Museum

NINETEEN

Creationist Makeovers

GENESIS OR GEOLOGY?

By the 1960s the scene had shifted again. The shock of America being beaten into space by the Russian launch of *Sputnik I,* the first artificial satellite ever placed in orbit, thrust the quality of our science versus their science into the hysteria of Cold War rhetoric. On the plus side, the *Sputnik* debacle at least prompted thinking about a renewal in public school science education. I know that I benefited from the boom in science education funding that followed. The creationists, once so loud, had vanished from the public eye in the years following the Scopes Trial, because they had for all practical purposes won and no longer needed to be active. Publishers had cooled creationist fervor by letting evolution slip away from school textbooks. Nevertheless, creationists lay like dormant termites within the walls of American life. When new curricula and high school biology texts eventually restored the teaching of evolution as a fundamental idea of biology, creationism reappeared in fully energized righteousness.

Creationists have three distinct but interrelated concerns about teaching evolution. The first is theological, the threat of destruction of the literal truth of the Bible. The second is that they believe evolution removes any purpose to life. The third is that they sense that belief in evolution is the root of all bad behavior, because it makes us "just animals" and so removes all grounds for ethics. The first worry is true but irrelevant to science, and hardly more damaging to literalism than is biblical criticism. The second concern is highly debatable, and the third is fatuous silliness. These fundamentalist frets fed the outrage that drove the twists of strategy of creationist campaigns over the years. As the Supreme Court by the 1970s had swept away the banning of teaching evolution as a viable option, the creationists started evolving a new sales gambit. Their switch was to a stance that claimed they also did science, called "creation science." Creation science is an oxymoron. Simply, creation science is to biology as astrology is to astronomy (or as an erectile dysfunction drug advertisement is to Shakespeare). Groups promoting creation science asserted that the Bible correctly told all when read literally, and that the six days of creation were a scientifically valid account. They also had touching attachment to the famous flood of Noah, which they rhapsodized into an account for the entire geological record, with all the fossils having dropped to the bottom in the great flood in the order in which they are found in the rock record. Thus trilobites are not hugely more ancient than people. They just sank faster.

This version of things is called "young Earth creationism," a fantasy take on science in which dating based on modern nuclear physics is negated and trumped by a chronology based on the life spans of imaginary nomadic patriarchs and the world is literally about six thousand years old. This daydream is unsustainable almost from the first line of Genesis. Literalists can't grasp the fact that the two books of Genesis tell entirely contradictory stories. Perhaps they should consider looking at R. Crumb's *Book of Genesis Illustrated.* In illustrated format, the nonconformity of stories is painfully obvious. Two distinct accounts from two cultural traditions have been uneasily jammed together to make a poor fit.

Noah's flood is firmly dated by literalists to about 2350 BC, despite the fact that the Egyptians were at that time busily building pyramids in the desert and not being inconvenienced or even feeling slightly dampened by the flood that fundamentalists are confident covered their structures, their pharaohs, and indeed everyone under a roiling depth of over five miles of water. In the last few decades of the twentieth century, a whole genre of heavy-breathing documentaries was born about intrepid hunters and their successive and nearly successful searches for Noah's Ark among the crevasses and glaciers of an obscure Turkish mountain called Mount Ararat. It's all high drama. Petrified boat hulls are glimpsed peaking out from mountain glaciers in out-of-focus photographs, old-looking but undated pieces of wood are found on the mountain side and identified as parts of a ship's hull. Reproductions of pottery just like those said to have once been seen by searchers in the remains of the ark are lovingly shown. Strangely, those who found the ark somehow never actually went aboard and collected any of that pottery. Money is raised each time from the gullible for one expedition or another to go back onto Mount Ararat again to recover Noah's toothbrush and log book.

If you are curious as to how I can be so confident that the promoters of Noah's flood are deluded, there are small matters of physics, geology, and history that render it an impossible story. For example, it's easy to do a simple rough calculation. To have covered the entire Earth as claimed, the waters had to have been as deep as Mount Everest is high (8,848 meters or 29,029 feet). The oceans of the world contain 1.3 billion cubic kilometers of water and have an average depth of 3,790 meters (12,430 feet). Oceans cover 71 percent of the planet surface. The flood had to cover 100 percent. So we multiply 1.3 billion cubic kilometers × 1.4 (the inverse of 0.71) × 2.4 (the depth fraction 8,848/3,790 = 2.4) to get the excess volume. The rough answer is that the water of the flood would have had to add up to just over three times the current volume of the oceans. So somehow about two and a half times the amount of water present on the Earth's surface had to be produced and added to the existing volume of the seas by Noah's deluge. Then, after forty days and nights, all that excess water had to be whisked away again. Pre-scientific writers thought that great underground caverns might have released the water and then imagined that it returned to these giant hidey-holes. No such enormous water-filled caverns exist or can exist in the Earth's crust. Nor could all of the excess water have been evaporated into space. The ice caps contain only about 2 percent of the water in the oceans. It's not hiding there either.

The claims that all of our geologic record comes from the flood was shown false over a century and a half ago – in fact, before Darwin published *Origin of Species*. I'm not making fun of the ark story. It is a remarkable legend from before the earliest days of written literature. But it is a legend, not a factual account of a global flood. The flood never happened. Part of the ancestry of the biblical tale lies even deeper in the past in the ancient Babylonian *Epic of Gilgamesh,* which contains the roots of the flood story and connects the Bible creation stories to a wider growth of human culture. The creation stories from the dawn of civilization are made ridiculous not by science, but by those who try to make us believe that they should be read literally in modern times.

LET CREATIONISM BE CREATION SCIENCE

The creationist legal strategy of the late 1970s was to demand equal time laws for teaching "creation science" on a par with "evolution science." By 1980, drafts of such laws were appearing in state houses across the country, like those big green flies that show up at a country summer hog roast. Bills were passed by a few impressionable state legislatures.

Not all of my involvement with evolution has actually been about doing the science, because evolution long ago became a social issue in America. I was only beginning to work with evolutionary problems in 1980 when the creation science equal-time-law road show came to Indiana. I decided to appear at the legislative hearing as a citizen opposing the bill. The Indiana State House is a magnificent late-nineteenth-century building that contains polished slabs of Indiana limestone containing fossils, silent witness to our planet's deep past. I soon learned how things are done in state legislative hearings – a different homage to tradition. The witnesses in favor of the bill each had about half an hour to make their presentations, while each of us who testified against it was allowed an entire five minutes. The creationist witnesses provided a lavish coordinated show of the awesome power of creation science by bringing out the big guns, such as a talk by a former high school administrator and a cartoon film explaining how evolutionists think "Bossie the Cow" got carried out to sea and evolved into a whale. Finally a lesson on the legal reasons that creation science had to be given equal time was paraded by a Yale Law School–educated lawyer Wendell Bird, staff attorney for the Institute for Creation Research. This discourse was based on his law journal article as a student. His legal thesis was a bogus propaganda hash intended to impress nonlawyers.

Real scientists testifying against the bill were craftily neutralized in this hearing by having a creationist cardiologist from the Indiana University Medical School present a detailed argument based on the red protein of our blood, hemoglobin. This molecule has just over 140 amino acids in its sequence, and there are 20 different amino acids. The witness did some Wizard of Oz math to show that for the correct sequence to arise at random, the number of tries would exceed the number of atoms in the universe and could never take place. Lots of pictures of models of globin molecules and calculations of probabilities were paraded before the legislators, all quite impressive on the face of it. Unknown to the lay audience, it was all simply an intellectual sleight of hand joined to a meaningless calculation. He was advertised with the title of "Dr." in front of his name, but this presenter was, as is typical of creationist experts, not well versed in evolution. No evolutionary biologist has ever argued that complex molecular structures like proteins arose spontaneously as they are today. Their origins lie in evolution step by step under selection from smaller and simpler molecules. I left the hearing depressed, and I was surprised when the bill ultimately failed. I think that the rejection, though a happy one, was hardly due to the brilliant oratory of we five-minute science witnesses. It seemed to owe more to the good sense of most of the legislators, and to the testimony of some church groups that opposed the bill because it favored the views of a particular, narrow Christian sect. Such bills passed easily in some other states and resulted in legal challenges that inevitably led to highly publicized and decisive trials.

A few years later I was invited to do an hour's televised debate with a creationist on Indianapolis' Channel 8. I was to debate a former Indiana state senator named Donald Boyes, who by trade was a preacher. Much to my surprise, he dressed the part. The cut of his maroon jacket would have been in character for a stage production of *Inherit the Wind.* Fortunately, I had learned how to prepare for debates with creationists by reading accounts of other such public debates, which by that time littered the landscape from coast to coast like verbal popcorn. I was several times invited by Kentucky Moral Majority to come down to Louisville and debate various creationist champions brought to town to devour evolutionists before audiences bussed in en masse from Pentecostal churches all over the region. I declined to waste my time on those circuses, but the Channel 8 debate offered a fair moderator and a sane audience.

By this time, debates between creationists and scientists had evolved into a format of sorts in which the creationist threw out one "fact" after another proving the Earth is young or that the second law of thermodynamics is violated by evolution, or that bombardier beetles could not have evolved their chemical defenses. The scientist had to refute these "facts" or lose credibility. It's not hard to do if you are prepared for all the stuff that turns up in the creationist rummage bag, but still it's a defensive role from which no decisive scientific argument can be made, and no concept appears on stage for long enough for it to be explained. I enjoyed the give and take and did well. I was especially pleased to get Boyes to admit that he thought that the Catholic Church was a cult, not really a Christian religion. A few days later I got a grumpy letter from his wife, who thought I was both disrespectful and had lost any chance of salvation. I was similarly denounced once in a letter to the editor of the *Bloomington Herald Times* by a local minister named Oliver Rogers over my role in science textbook committees and public discussions of choice of biology textbooks for our high schools. He told me that he felt sorry for me because, in the certainty of his belief, he knew that I had "no purpose in living, and no hope in dying." People born and raised in modern America in an ostensibly shared culture in fact can be separated by a centuries-wide intellectual crevasse.

The federal courts would become the most effective means of defeating the notion of equal time for creation science. There, expert witnesses representing both sides could be examined and cross-examined under oath by the trial attorneys. The first challenge to the 1980 creation laws was not long in appearing. In 1981 the Arkansas version of a creation science act was challenged in U.S. District Court in a case called *McLean v. Arkansas Board of Education.* The judge was William R. Overton, who unlike Judge Raulston of Scopes fame had a sense of curiosity and admitted expert testimony to establish whether the act in question represented science or religion posing as science. Overton wrote an insightful decision in which he went to the heart of what science is. He found that evolution fits and creation science doesn't. As district court decisions are limited in geographical extent, Overton's decision didn't affect things in other states, but that was coming. In 1987, the U.S. Supreme Court heard the case of *Edwards v. Aguillard,* which arose in Louisiana. The justices ruled against creationism in a seven-to-two decision. The critical issue was that the law served no secular purpose. Its aim was to have public schools promote religion. The fact that there were two Supreme Court justices who voted in favor of the creationist charade was the only bizarre feature of the ruling. One of the two justices who voted in favor of teaching creation was Antonin Scalia. As recently as an interview in 2009, Justice Scalia said that he is against separation of church and state because "we are a religious people whose institutions presuppose a Supreme Being."

INTELLIGENT DESIGNERS

These cases should have ended the matter in the United States, but creationism is like the Phoenix of legend. Declare it dead as science, then stand back and watch it arise from the ashes and assume a new guise. Creationism was thus once more reborn a few years ago in the resplendent plumage of Intelligent Design (ID). It helps to realize that the idea of design in nature is not new. This is a venerable idea that goes back to Aristotle and was enormously popular at the time Darwin was a student. Darwin read about design in the form of the influential book *Natural Theology*, published by William Paley in 1802. Here the analogy was made between a machine, Paley's legendary watch lying on the ground, and an organism with its appearance of design and purpose. Paley lovingly detailed such elements as the human eye, so perfectly designed for sight. The appeal of design to humans is obvious. When we create something, we operate in a top-down way, like engineers. When we make something, we think of a design and execute it using the appropriate tools. Even if we are simply tinkering with a preexisting device, we have a notion of what we would like to achieve. Machines are built with a purpose in mind and are defined by the function they serve.

Living organisms are intricate and carry out "functions." Their parts seem so well designed to match the function they serve that we have a hard time contemplating them in any other way than as designed. Even language gets in our way if we try to express the idea that organisms fundamentally are not watches or bicycles. We know what lungs or eyes or legs are "for." We say that body structures appear "well designed" for what they do. There are a few nasty problems with applying the idea of design to biology. Is an organism a machine? Who is the designer? Does the designer have to be intelligent? What about the problem of infinite regress? That is, if complex entities need a designer, then our designer had to have been designed by another, but superior designer, and so on up the chain of designer creation.

Natural selection is different; it is a bottom-up mechanism. There lies the naked ugly truth that has made natural selection unpalatable to so many. There is no benevolent designer, just trial and error, just failure and success in reproduction. There is no goal in evolution, and no divine guidance. Natural selection differentiates between those individuals that are more effective from those that are less so. Selection cannot look ahead, nor can it produce perfect organisms. To avoid providing lunch for the wolves, an individual reindeer doesn't have to be the fastest runner in the herd, just faster than the slowest. For most people, this is not a picture of a benevolent world. Despite our experiences in the real world, earthquakes, floods, plagues, and starvation, we can't quite believe it. The contingency of events is disquieting. The evident tragic unfairness of mindless contingency to some extent mitigates the problem of a god who allows evil, which has so severely tested the faith of people who have lost a child or have watched the greedy thrive at their expense. One might regard parasites such as tapeworms as incomprehensible evil, but the role of selection in evolution gives us the answer to the problem of why the majority of all animal species are parasites, which use other species to their own ends and often bring gruesome pointless misery to their hosts. Being a parasite works ever so well. That's just the way it came out. No god is consciously meting out evil as a test of faith.

Intelligent Design is interesting because it is not another Bible-based faux science put on by the usual cast of bumpkin creationists. Its advocates are sophisticated and well educated, including among their leaders a retired professor of law at Berkeley, a follower of Reverend Moon with a Ph.D. in developmental biology from Berkeley, another with a Ph.D. in paleontology from Harvard, a professor of biochemistry at Lehigh University, and a professor of philosophy at Baylor University who has since relocated to Southwestern Baptist Theological Seminary in Fort Worth, Texas. With a team like this, surely there is fire somewhere under all that smoke. Not really, just well-educated smoke. The Discovery Institute in Seattle, which is the home base of ID, produces no actual scientific research and actually discovers nothing. What they promote is lots of propaganda about ID science, but their real aims were revealed when one of their internal documents, a sales strategy that has become known as the "Wedge document," was leaked. I downloaded it from the National Center for Science Education website. This section describes the three phases of the project, summarized briefly:

Phase I. is to do the vital "research at the sites most likely to crack the materialist edifice."

Phase II. "The primary purpose of Phase II is to prepare the popular reception of our ideas. We intend these to encourage and equip believers with new scientific evidences that support the faith, as well as to 'popularize' our ideas in the broader culture."

Phase III. "Once our research and writing have had time to mature, and the public prepared for the reception of design theory, we will move toward direct confrontation with the advocates of materialist science."

The propagandistic nature of this Wedge plan is plainly revealed in that Phase III is planned before the "scientific" results of Phase I become known. In a real scientific revolution, ambiguities in current theory would prompt proposal of new hypotheses. These would generate experimental tests. The evaluation of theories in conflict would be decided by scientists, not the pope, opinion makers, or high school students. The Wedge document doesn't discuss the science to be done. The object of the Wedge program is not to persuade scientists but to influence "opinion makers" and ID's "natural constituency, namely Christians." In real science the goal would be to find a better explanation for natural phenomena. The document reveals the goal of the ID movement is "to replace materialistic explanations with the theistic understanding that nature and human beings are created by God." Frank Ravitch, in *Marketing Intelligent Design,* points out that "from the perspective of the religious apologist, the end of serving ultimate truth justifies the means." The Wedge program is not science. It is propaganda, and it is propaganda funded in part by a wealthy Christian reconstructionist,

Howard Ahmanson Jr. It is his privilege to donate to whom he wishes, and the Discovery Institute is free to accept his support, but doing so shows that they are only pretending to be scientists.

Perhaps the ID approach sounds harmless, but not once we have seen the results when dubious philosophies are applied to public policy. One highly publicized instance took place when Ronald Reagan appointed James Watt as secretary of the interior. Watt did not believe that any conservation of nature or resources was important because Jesus would be returning soon. Mining and lumber companies could prosper happily in the meantime. Reagan was also notorious for acting, or more accurately, not acting, on his belief that AIDS was a punishment meted out by God on homosexuals. He thus opposed and delayed funding for research and public health measures. AIDS is caused by a virus that was acquired by humans infected as a result of hunting of primates in Africa. AIDS is sexually transmitted and doesn't care about whether its victims are homosexual or heterosexual. It is tragic, but not supernatural. Reagan's spiritual descendant, the second George Bush, institutionalized fundamentalist theology as a part of the workings of government agencies dealing with birth control, disease prevention, stem cell research, global warming, and energy policy. The effects were just as heartbreaking as could be expected. Scientific panels were packed by ideologues. Irreplaceable years were lost on dealing with what could become irreversible environmental and population crises. Global warming was pushed to the back burner, where it could be ignored. Yes, it might be possible to pay no heed to a teapot full of gasoline on the back burner, too, but in either case the outcome won't be pretty. Ignored, disbelieved, and laid at God's door as political whim dictates, but finally we will own the disaster.

In a pattern of dysfunctional behavior as familiar as compulsive gambling, attempts by school boards to insert creationism in the guise of Intelligent Design into science classes, as a scientific alternative to evolution, became a new industry in America. Inevitably, another court case was soon to follow. This one, *Tammy Kitzmiller, et al. v. Dover Area School District, et al.,* was decided in December 2005. The motivations of the creationist members of a small Pennsylvania town were echoed in a sermon by a local minister, who said, "We have been attacked by the intelligent, educated segment of our culture." This is a book burner's take on intellect. Did he think that science is a religious cult devised by the "educated segment" to seduce children into evil with pictures of godless geologic timescales?

Dover may be the most interesting and well-documented creationism trial so far. At the trial, Judge John Jones admitted testimony from both evolutionary biologists and ID scientists. The testimony of long-suffering teachers and devious school board members would be riveting. At least four readable and well-informed books were published about the saga. It can only be the greatest of ironies that the most amusing account was written by Darwin's great great grandson, Matthew Chapman, a moviemaker by profession. Chapman came to town for the trial and was entranced enough to hang on through the entire event. He called his book *Forty Days and Forty Nights* (after the fortuitous exact length of the trial). Trials produce revealing exchanges; here an attorney for the school board questions a witness for the plaintiffs, John Haught, a Catholic theologian:

> "Intelligent design is different than creationism, is it not?"
> "Yes, in the same sense that, say, an orange is different from a naval orange."

The contest did not go well for ID. It was discovered that the book that the school board had tried to adopt as a supplemental ID text, *Of Pandas and People,* had been drafted originally as a creation science text. Where those words appeared, they had simply been replaced by the words "Intelligent Design." ID science witnesses dragged out impressive-sounding cellular structures as examples of entities too complex to have evolved. The all-time favorites, the bacterial flagellum and the enzymes of the blood-clotting system, were thrust into the fray as unbeatable lions, irrefutable evidence of design. They proved to be easily disposable sacrificial lambs. The ID examples lost their glow because research had by then shown that they in fact are not irreducibly complex and that their simpler evolutionary precursors actually do exist and function in nature – just as predicted by evolution.

How inconvenient the continuing discoveries of science can be. In a 2009 publication molecular evolutionist Russell Doolittle would further demonstrate just how completely the supposedly irreducibly complex blood-clotting system is rooted in the long history of vertebrate evolution. The most primitive jawless fish have the genes for the thrombin-catalyzed conversion of the precursor fibrinogen to the clotting protein fibrin, but these creatures lack several supposedly irreducibly complex clotting factors. It is not until one reaches the pouched marsupial mammals that the full complement of proteins of the human system has evolved.

After days of listening to arguments in support of ID as science being laid to rest, the trial closed. A few weeks later, the judge's ruled that ID was religion and not science, and he took time to note that the actions of the school board were a "breathtaking inanity." He was just in time to get a few death threats for Christmas. There is an amusing footnote on the part of the attorneys involved in the suit against the Dover school board. Two of them, Stephen Harvey and Eric Rothschild, record that the standing joke during the trial was that "creationists are the best evidence for evolution: they adapt to a hostile legal environment."

SCHOOL BOARDS, SO EASILY SUBVERTED

With this last failure in court to ring in the new century, creationism has moved on to a strategy of requiring that schools feature "critical thinking" and "teaching the controversy." In that dodge, teachers are expected to point out the flaws in evolution so that students can evaluate the validity of alternate scientific views in evolutionary biology (namely, Intelligent Design). Such latitude doesn't seem to extend beyond evolution. No one asks ninth graders to decide on alternate views of numbers theory, chemistry, quantum mechanics, or black holes. Rather than fighting court decisions, creationists now focus largely on packing and subverting school boards. The most spectacular example was the recent fight in the Texas State Board of Education. In 2009, the Texas school board was freshly loaded up with creationists and chaired by a creationist dentist appointed by a governor running for reelection and seeking to score points with religious conservatives. The national importance of the move to influence science standards in Texas into an anti-evolution stance is that Texas has so large a share of the school textbook market that its tastes heavily influence what is included in textbooks everywhere in the country. Nonsense decided upon in Texas will end up in school texts used even in districts free of weekend cowboys. This battle reminded some of what Mark Twain observed in 1897: "In the first place God made idiots. This was for practice. Then he made school boards."

This same Texas state school board in 2010 shifted its efforts to the state standards for teaching American history. They aimed at turning the Founding Fathers from the thoughtful deists that most of them were into fiery-eyed evangelicals. By March, the Texas board had voted to remove Thomas Jefferson from American history texts because he was a subversive leftie suffering from insufficient religious fervor. But we make fun of the efforts of creationists and history revisionist at our peril. These ideas are cuckoo's eggs laid in the nest of democracy. All activity on the part of history revisionists I think comes from their realization that our past creates the setting for our present. They understand that to control the present, it is critical to control the past, and to control the past means changing our conception of it to suit ones political or religious beliefs. George Orwell in his novel *Nineteen Eighty-Four* created a frightening society in which totalitarian rulers effectively ruled by a day-to-day remolding of the past, as was formerly done in the now-extinct Soviet Union. Observers used to watch for the removal of high Soviet officials by looking at doctored photographs of the previous year's May Day parade to learn who had been airbrushed out of last year's official picture. The study of evolution impedes the control of the deep past by those who would impose their own view of the world and humanity.

In line with the supposed ancient Chinese curse "May you live in interesting times," I've watched the continuing, and very interesting, rise in popularity of creationism in the guises of creation science and Intelligent Design. America seems particularly vulnerable to creationism – in part because our television media believes that all ideas are equal and thus always gives us two opposed talking heads on every issue. A scientist and a creationist are given equal time. This procedure gives crank ideas an undeserved respect. Critical thinking and knowledge are not required in these contests. The confused public is left to decide on a complex question from a five-minute face-off muddle. Supposedly the result represents balance, with both sides represented. The fact that one side is gaseous nonsense escapes notice in the shouting.

Science education standards in many states require that evolution must be presented. However, in many communities hostility from parents or a creationist teacher renders the requirements ineffective. The dominance of conservative religious outlooks has an enormous effect. In 2005 a survey of the public acceptance of evolution in thirty-four advanced countries placed the United States at thirty-third, with only Turkey making a worse showing. Even Croatia, Romania, Bulgaria, and Slovenia do a better job teaching evolution than we do. Perhaps our pride can be assuaged by the fact that creationism is even more prevalent in the Muslim world. A worldwide survey shows that it's only in Europe that a majority of Christians accept evolution; North America, South America, Asia, and Australia are all not so good. We are not alone in obtuseness.

The creation of a bogus but passionately felt connection between a particular scientific theory and evil is disturbingly medieval. The actor Ben Stein, best known for his minor part in the teen movie comedy *Ferris Buhler's Day Off,* more recently produced a so-called documentary, *Expelled,* in which the "persecution" of ID science by the mainstream is revealed – with a big stretch of the evidence. In a 2008 television interview on a the fundamentalist religious Trinity Broadcasting Network, Stein said, "When we just saw that man, I think it was Mr. Myers [biologist and blogger, Dr. P. Z Myers], talking about how great scientists were, I was thinking to myself the last time any of my relatives saw scientists telling them what to do they were telling them to go to the showers to get gassed. . . . That was horrifying beyond words, and that's where science – in my opinion, this is just an opinion – that's where science leads you." No, not quite. A comment like this shows where willful ignorance leads you. Stein doesn't seem to know, or perhaps doesn't care, that the first president of Israel was Dr. Chaim Weizmann, a chemist, who also founded the Weizmann Institute of Science in Israel. Sadly, he doesn't need to know, because the Trinity Broadcast Network by 2009 was the largest broadcast network in the world and capable of creating a reality of its own.

Such efforts should not be dismissed lightly. They have influence, and that influence can sometimes lie beyond popular belief in realms where it should have no impact at all. Perhaps the most shocking example is that the U.S. National Science Foundation's 2010 edition of *Science and Engineering Indicators,* which is meant to keep track of American's science literacy, omits any mention of evolution. Specifically, two survey questions and their responses were deleted. These were "Human beings as we know them today, developed from earlier species of animals" (45% responded true) and "The universe began with a huge explosion"

(33% said true to this). NSF officials said that they removed the items because they felt that the questions were "flawed indicators of scientific knowledge because the responses conflated knowledge and beliefs." Am I being slow, or isn't the lack of understanding of the difference between science and religion betrayed in the responses a significant indicator of scientific literacy? The questions asked were not just matters of opinion.

MEANWHILE, BACK ABOARD THE ARK

If you like willful ignorance and its sometime silly outcomes, American school conflicts provide a rich theater of unending farces. In 2009 the *Springfield News-Leader* reported that the Sedalia, Missouri, school superintendent had ordered the school band to recall its new t-shirt. The band's theme was "Brass Evolutions," and the shirt made clever use of the classic lineup of monkey to ape to hominid to humans marching along, but each with a trumpet in hand. One band parent who teaches in the district is quoted as saying, "I don't think evolution should be associated with our school." A sufficiency of similarly enlightened parents complained about what they saw as the evolution theme. The superintendent ruled that the shirt had to go because the "district is required by law to remain neutral on religion." So even a mention of evolution is religion? What about physics, or organic chemistry, or anatomy, or astronomy? Would they count as religion too?

Another such debacle was reported in a 2006 issue of the Waco, Texas, *Tribune Herald*. Bill Nye, famous as "Bill Nye the Science Guy" for his excellent television science series for children, was invited to give a lecture and receive an award from a local college. Nye mentioned celebrated lines from Genesis: "God made two great lights – the greater light to govern the day, and the lesser light to govern the night. He also made the stars" He went on to say that we now know that it is the Sun that produces the light and the Moon only reflects it. Some members of the audience were angered by this observation and left mid-program. One woman who had hurried her three children away from such planetary depravity was quoted as saying, "We believe in a God!" Mind you, this Texas dustup over the light-producing capabilities of the Sun and Moon took place four hundred years after the facts became clear to astronomers. Anyone who remembers the mid-twentieth-century comic strip *Pogo Possum* remembers Pogo's lament: "We have met the enemy and he is us."

I guess we can at least be proud of being just ahead of Turkey in acceptance of evolution. But another look suggests this is only a part of a parallel bad trend. Turkey is a majority Muslim country that established a secular state and constitution following World War I and the collapse of the Ottoman Empire. Muslim fundamentalists now vigorously challenge the secular status of Turkey. A key wedge issue for them is the forcing of creationism into Turkish schools. The effort has succeeded in suppressing most opponents in Turkey. A driving proponent of this mischief is a wealthy character named Adnan Oktar, who (evidently with swarms of anonymous ghost writer helpers) creates a flood of creationist material under the pen name of Harun Yahya. That he feels he needs a pen name when we all know his real name is just one of his foibles.

Yahya's willingness to spend large sums of money was dramatically illustrated in 2006, when I and many of my colleagues received gratis a twelve-pound green book from him titled *The Atlas of Creation*. This 800-page tome is lavishly produced and packed with wonderful color photographs, many of superb fossils. The text is not so wonderful. The basic strategy of the book is to show pictures of fossils similar to living forms and to claim that this similarity proves that no evolution has taken place in insects, salamanders, birds, plants, or anything else. Yahya is notably catholic in his coverage and tediously repetitive in his simplistic message. I'm puzzled that he would have bothered to go through the expense and trouble to send this expensive tome to so many evolutionary biologists, who would find it an amazing curiosity but never scientifically convincing. But it can still be a successful effort despite its scientific fallacies if sent to sympathetic nonscientist audiences. The gains for creationism in Turkey are disheartening. As it is, evolution is poorly accepted in most Muslim countries. Wherever they occur, the spread of the retrogressive ideologies of fundamentalism in major religions can only retard the development of more rational societies.

I address the religious origins of creationism directly in my undergraduate classes in evolution and explain the distinction between science and religion. Science can be defined as the seeking of natural explanations for the phenomena of the natural world. Thus we can discover from the application of science the basic facts about our world, such as that the Earth is unnervingly ancient and that we have even been able to measure its age. The success of scientific explanations is borne out by the effectiveness of the hypotheses made. If the hypotheses fail the test, they are dropped from further consideration. Correspondingly, religious views are untestable and so cannot be admitted into science. In general, public education does a poor job teaching the notion that science can only be studied if we realize that nature is the result of natural phenomena that are underlain by natural laws. What creationists, including the Intelligent Designers, want is to insert supernatural explanations into science on an equal or even superior footing to natural laws. These explanations exist in the Bible, why shouldn't they be a part of studying the natural world? The answer is simply that once supernatural causes are admitted, they slowly strangle the impulse to do science. First, they make a joke of the replication and regularity of natural laws. Second, they admit hypotheses that can't be falsified. Third, they remove the very motivation to ask questions. If something is hard to explain, let's not bother about the effort of seeking a natural cause and just leave it as an expression of a divine plan or even God's whim. Science is hard, but religion is easy.

There is a clear application of the concept of natural phenomena arising as a result of natural processes. People have been highly tempted throughout history to see natural disasters as "acts of God" administered as punishment of some communal sin, and to seek scapegoats to punish in expiation. But as these disasters are understandable as the results of discernable natural laws, we can predict with certainty that they will occur again and again, whether we are naughty or nice, no matter what theological dogma we believe. We can also know why, and thus how, disasters might be predicted and ameliorated.

DINOSAURS IN EDEN

There are dinosaurs aplenty in our culture. Children grow up with them as virtual playmates. Now dinosaurs have a sacred job, too, and have been recruited as companions for Adam and shipmates for Noah. There is of all things a newly established and well-funded Creation Museum in Kentucky that features dinosaurs frolicking with Adam and Eve. By the theology of this museum, there was no death before the Fall, so no fossils can predate Eve's fateful rendezvous with the serpent. That being the case, most fossils became fossils by being drowned by the flood. The now-extinct dinosaurs had to have ridden out the flood as passengers on Noah's ark as required by the story in which Noah takes a pair of all beasts aboard for the great sea cruise. The museum has a replica *Triceratops* strapped into a saddle. I know it's used for children's pictures, but I enjoy thinking about it as a representation of the very dinosaur Noah rode while he herded the other dinosaurs onto the ark. A reporter revealed that the full-sized replica of Adam was embarrassingly based on a local porn star "hunk." This gentleman apparently had not been well screened before he was invited to serve as the model for the father of humanity. Even more strangely, another observer noticed that the Adam statue lacks a penis. The omission of this awkward bit of anatomy may be "family friendly," but honestly, it is unrealistic in a garden where the inhabitants were instructed by the Creator to let it all hang out. As the text says, "They were not ashamed." That account was joined to the well-known admonition in Genesis to "be fruitful and multiply." Comedian Lewis Black said of people who attend the place as an educational experience, "These people are watching the Flintstones as if it was a documentary."

I want to note, though, that the Creation Museum does present a sketch of their philosophy of science, and it's an interesting one. Their concept seems to have three parts. They accept the accuracy of what they call "operational science," how things work in the present, but they claim that science cannot investigate the deep past. Instead, it's just a question of looking at evidence such as a dinosaur bone from a different perspective, which gives an equally valid different interpretation. The whole museum proclaims the third point. The only perspective from which we can really understand "origins" is through the genuine written words of the Creator that tell how he did it – literally and conveniently recorded in the Bible, of course.

Where do the crazy images come from? Dinosaurs are immensely popular in advertising, on science fiction programs, and now even as playmates on rainy days for Noah's children. Plastic dinosaurs are for sale everywhere, but there is not so much saturation in teaching the actual science of how we know that dinosaurs ever existed and when. Students are rarely well taught before college (if even then) how scientists have come to know that there was an evolutionary history of life and how the physics of decay of radioactive isotopes led to the discovery that the Earth is billions of years old. The result is that nearly half of Americans think the world is actually only six thousand years old and that the age derived by nuclear physics is just a guess by evolutionists. Sixty percent of Texans surveyed in 2010 reported that they thought dinosaurs and humans co-existed, or they were not sure. Similar numbers believe that humans were created just as we see them now – except that we now have better dentistry, deodorants, hair dressers, tanning salons, botox injections, breast implants, plastic surgery, and hair restoratives than those available to Eve after leaving the Garden or from the cosmetics bar on the recreation deck of the ark.

I don't want to leave anyone with the idea that an absolute dichotomy with divine creationism and atheistic evolution as the only possibilities. Surveys show that a substantial fraction of people, including some scientists, accepts versions of theistic evolution that are consistent with the practice of science. Theistic evolution ranges from a direct role for God in directing the course of evolution to a deistic view that God set up the basic rules of the physical universe and has since then stood back to allow evolutionary processes to take their course. Theistic evolution and deism are religious and philosophical, not scientific, ideas, but they allow people to rationalize the science of evolution with personal religious belief. Others have taken the view that science and religion simply operate in different spheres. Again, this is a view that allows people to accept evolution without having to accept the kind of dualism that creationism requires, where one is limited to being either a godless supporter of evolution or a biblical literalist.

There is no law that declares the human species to be immortal.

Richard E. Leakey

I'm not a scientist. That's why I don't want to deal with global warming.

U.S. Supreme Court Justice Antonin Scalia

TWENTY

Evolution Matters

CURIOSITY

If there are two common paths for children to become entranced with science, either through an early interest in nature or through a later intellectual awakening in school, there seems as well to be another kind of division that comes into the kind of science we do. For some scientists the draw is basic science, but for others the pull is toward research that has an application in mind. This division cuts across disciplines. There is applied ecology and paleontology just as there is applied molecular biology. The utility of science to human needs was built into the origins of modern science. The link between scientific anatomy and medicine goes back to the remarkable sixteenth-century anatomist Vesalius. Although first-class science does not have to be motivated by practical or humanitarian goals, it many times is. Important research has been directed at the discovery of cures for diseases, the discovery of new materials, or the development of useful technologies, including recently fiber optics, digital photography, and computers, all of which make major contributions to the quality of our lives. In my own life growing up, I watched my father devote his career to designing new polymer materials for useful applications. His view was that science had to have a practical expression, and it was doing that science that he found compellingly interesting. I was influenced by my father to be drawn into science, but I could never feel any appeal for applications of science that attracted him.

I do science because I'm curious about nature. For me a life in science has been a purely personal thing that grew out of my childhood life as a naturalist. I was not driven by any motivation to do practical, good (or bad) works. My aim has been to solely to satisfy my curiosity and to help build human understanding of the world. I know that my simple personal motivation may seem selfish, but I make no apologies for that. Lives dedicated to scientific curiosity have shown us how our cosmos has evolved and how it functions. The public image of scientists casts us as serious, white-coated workaholics with little humility and with not much in the way of human feelings and joys. Old sci-fi movies couldn't imagine a scientist who wasn't either an evil villain or the impractical and nerdy character scripted to be eaten by those giant ants. In real life, not a few of us enjoy science as intellectual play and discovery as a form of self-expression and creativity as profound and rewarding as creating literature, art, or music. To make a discovery, no matter how small, is to see and understand something that no other human has known before and to introduce something new to human knowledge. It is not to read what has been written down earlier, but to reveal and understand something new, something that has not been recorded by anyone else. Discoveries are not merely the addition of new facts. They affect how we know, and they can overturn even favorite theories. Rather than being an end in themselves, discoveries are the crucial steps in the process of understanding. Most of the creativity of science comes from designing a way get to the moment of discovery. Discovery lies at the heart of what compels the community of science, and discoveries made under the banner of simple curiosity have transformed the world.

The search for human ancestors in the fossil record by Louis Leakey and his successors is immensely inspiring in understanding how humanity arose, but it has no practical meaning by the usual commercial or military definitions. Darwin never mentions any business plan he hoped would develop from his studies of evolution. To Darwin, orchid pollination, coral reefs, human emotions, and the humble unseen work of earthworms were all simply beautiful puzzles of nature to be solved. The process of finding the solution was to be enjoyed. Yet basic science, in this case evolution, has transformed how we understand the universe and our place in it – thus the reason we hunt for extrasolar "earths" is to gain knowledge that will help us understand the evolution of our own planet, as there is no practical hope of ever visiting or exploiting such far away real estate. As we so well know from the endless controversy over teaching evolution, what we find about our place in the universe is not always pleasing to our religious certainties, to our egos, or to our sense of importance in the cosmic sense.

With time I've learned that the tension between basic and applied is not so straightforward as the ideal of pure science might suggest, no matter the purity of our intents or the scholarly nature of our interests. Even when curiosity alone has been the goal, there also is an objective value in even the "impractical" discoveries of basic science. The objective value important to us now is our ability to assess how humans are doing as guardians of their planet. Various disciplines let us track the health of ecosystems and pressures on living populations. Developmental and evolutionary biology show us our own place in the animal kingdom and teach us that we are as vulnerable to the poisoning of our environment as our animal relatives. Evolutionary biology and paleontology give an understanding of the forces that drive extinction and warn us about how we might ourselves push life over the edge. These disciplines also show us how long the recovery of diversity is from mass extinction. On our time scale, extinction is forever.

What I want to address in this last chapter are some of the social issues that arise from bending science to advance aims that may be partially informed by science but have narrow objectives that conflict with a broader intelligent application of science. It seems that in an age of science, all science can be misapplied. Equally important, ignorance and misunderstanding are as dangerous as deliberate misapplication and political obfuscation for private gain. In the modern world, the problem arises especially strongly with evolution, which would seem at first sight a completely impractical sort of subject. Creationism helps confuse the public about science and reduces public trust in scientists who speak out on issues where accurate scientific knowledge is vital to reach intelligent solutions. Evolution is going on around us as a result of our impact on the Earth, and often not in our favor. Our growing environmental crisis has become our most pressing challenge. Evolution matters, and we need to talk about why.

IGNORING EVOLUTION

Our choices are not just between doing basic or applied science. All science comes to have a societal effect, and in turn social forces complicate any simple separation of basic and applied science, even the concept of societal good. The importance of the proper application of science can't be better shown than by the lives saved by vaccination against disease. The long campaign of smallpox vaccination helped lead to the extinction of one of history's great killers in the human population, but sadly and foolishly, the last surviving smallpox virus is still maintained in freezers in the United States and Russia. The number of vaccinated people is falling rapidly, and the virus thus becomes ever more effective as a weapon in waiting.

Another murky side of the relationship between basic science and dangerous applications was unveiled for me a number of years ago by one of my Indiana University colleagues, Martha Crouch, a developmental biologist who did basic science aimed at finding out how plant embryos give rise to a plant. Marti came by my office one day to tell me that she was leaving research. I knew Marti as an enthusiastic and successful scientist, so her announcement came as a complete surprise. She had observed that agribusiness corporations were becoming interested in plant development and genetics, and she realized what they intend to do with that knowledge. Corporations had already started using her research and other findings to help modify oil palms. Monoculture plantations of these trees were displacing great swaths of high-diversity rain forest. In addition, chemical companies were starting to buy up seed companies. That seemed like a surprising move for chemical companies, but Marti was right about where they were heading.

What the corporations intended to do was to control what seeds could be planted. The now-familiar story of corporations genetically engineering plants to make them more resistant to commercial herbicides so that more herbicides could be applied to crops had begun. In other words, the knowledge of plant genes and development would not be used to create plants whose farming produced less chemical contamination, just the opposite. The desired goal was to create crop plants that could withstand the herbicides produced by the companies that also marketed the resistant plants. Marti felt that no matter how basic the aims of her research or how well intentioned, it would end up contributing to an environmentally destructive commercialization that used plant biotechnology to promote more, not less, use of and contamination by agricultural chemicals. More pounds of chemicals sold, more dollars taken in. No need to worry about where those toxic pounds go or what they pollute and poison downstream.

Marti's predictions were unfortunately right on target, and ultimately not limited to oil palm or indeed to any one crop. All crops are "genetically modified," generally through classic methods of crop breeding over many generations. It's not the technique of genetic modification of plants that's critical. What matters is how it's used. Bioengineered crops are widely accepted in the United States and banned in Europe, without in either case a clear public understanding of the benefits and risks. Genetic modification potentially offers a great deal in the way of improving the nutritional value of crops, improving yield, and even allowing safe pest control by expression by the plant. For example, insect-killing toxins can be made in roots attacked by pests. This kind of modification promises to reduce the application of pesticides and thus be environmentally beneficial. This approach is widely used but unfortunately has its drawbacks. Fatefully, the chemical companies discovered that genetically modified crops could be used in a highly profitable way to sell more chemicals along with modified strains of crop plants that could tolerate levels of herbicides toxic to weeds. Thus was born Monsanto's mighty herbicide Roundup. As the seeds for herbicide-resistant crops themselves belong to the companies, farmers are forced to buy their seeds for each year's crop, as well as the herbicides, from the same company. This is the way one of the most creative applications of genetic modification of crops has gone.

Roundup, the trade name for glyphosate, became an enormously profitable product. It is widely used by farmers, who buy both the herbicide and "Roundup-ready" soybean and corn seeds from Monsanto. The herbicide can be applied to the resistant crop, and it simultaneously prevents nonresistant weeds from growing. How does that work? By clever gene engineering. The bacterial version of a gene that encodes a key protein, an enzyme required for the synthesis of an amino acid vital to plant cells, was modified. The mutated gene was introduced into crop plants. Now the crop plants carried a resistance mutation to glyphosate derived from the mutated microbial gene. The mutation prevents glyphosate from binding to the active site of the enzyme, and so the herbicide is toxic to weeds that lack the mutated version of the gene but not to the modified crop plant. The concept worked wonderfully, and the amount of herbicide needed dropped in the first years of widespread use. Then a little problem emerged from the ground. Glyphosate-resistant weeds sprang up in thick, unmanageable tangles. The result has been an enormous increase in use of other herbicides to kill the resistant weeds, trapping farmers on what's been called the "herbicide treadmill," the use of higher and higher levels of Roundup – and when that fails, more of the older weed killing chemicals as well. Monsanto is now providing the other weed killers at a discount to keep farmers hooked on Roundup-ready seeds.

The resistance came as a surprise – but it shouldn't have. When evolution is ignored and simplistic approaches to chemical use in agriculture are taken, things go wrong in ways that continually make Cassandras out of evolutionary biologists, who after being ignored when they could have made a difference are able only to watch the disaster unfold when it's too late. The story of Roundup repeats a familiar pattern, evolution under selection. It's the old story of resistance arising from attempts to control insects with pesticides or disease bacteria with antibiotics. The evolution of resistance to antibiotics is an enormous health menace, with our best weapons being rendered useless one after another. This drama has been played out repeatedly over the past fifty years, and it always has the same outcome. How about the environmentally friendly genetically modified corn that produces the natural insecticidal toxin Bt, which kills rootworm with no need for added insecticide? This application of genetic engineering has been highly successful, but here too natural selection will not be denied. Where not planted appropriately, insect pests rapidly evolve resistance.

Evolution acts through natural selection. In the case of Roundup, the poisoning of weeds was powerful selection for spontaneous mutations that gave the plant resistance to glyphosate. The result is evolution of weeds that now cannot be poisoned by modest levels of Roundup.

Some mutations affect the binding site for glyphosate, or modify a different part of the protein, or block entry of the herbicide into plant cells, or make many more copies of the gene, allowing vastly increased production of the enzyme to simply swamp out the herbicide. Different mutations are selected for as they arise, because these mutations allow plants that carry them to survive glyphosate. No one mutation is necessarily the "best." Good enough to survive glyphosate is good enough. Under selection, resistance evolves – inevitably.

The seemingly inescapable misapplication of science lies in wait in many ways, and modern military and commercial enterprises are well able to afford to buy any scientific expertise they desire. The story of one molecule can embody grotesque misapplication at work. Atrazine is a highly profitable and widely used herbicide on farms in the United States, but it doesn't stay down on the farm. Atrazine contaminates water all over the Midwest. In my state of Indiana, contamination levels of public water systems are commonly reported to be as high as 10 parts per billion, and to have spiked to 275 parts per billion. Tyrone Hayes, a biologist at Berkeley interested in how hormones control development in frogs, has been the chief investigator of the effects of atrazine. He observed the effects of atrazine in nature, and in facing up to his responsibility to address a dangerous misapplication of science, he went public with his findings. Hayes discovered that atrazine caused developing male frogs to become feminized, even to the point of being able to mate with normal males and produce viable eggs. The level of atrazine required is low – 0.1 parts per billion is sufficient. That is a level thirtyfold below the 3 parts per billion declared safe by the Environmental Protection Agency.

Okay, we might think that sterility and becoming abnormal intersexes is sad for the frogs, but so what? Except there is an evolutionary hitch in trying to take a disinterested standoff view in which we think that humans are somehow endowed with a body distinct from and superior to those of our evolutionary relatives. The truth is that all of us vertebrates use the same sex hormones, and atrazine has the same effect on all vertebrates tested, including mammals, which means our children and grandchildren, too. The sinister effects of atrazine are recognized in Europe, and use of atrazine is banned there. However, each year millions of pounds of it are legally used in the United States. The effects of agrichemicals on human birth defects are now emerging. Babies in agricultural areas show increased birth defects if they are conceived between April and July, when herbicides are being applied. The science is clear, but policy is unaffected. Corporate profits trump science, common sense, and love of our children.

THE EXTINCTION MACHINE

The examples I've just mentioned illustrate how our failure to appreciate the power of microevolution, as in the fast evolution of herbicide resistance, can threaten our well-being. You might wonder, though, why it might be important to give a thought to macroevolution beyond seeking the amusement value of dinosaurs. First, understanding macroevolution gives us an appreciation of where we came from and so helps free us from irrational explanations of our origins. It helps focus our attention to the earth as our long-term home and to the dangers of human actions that disrupt our living conditions through the heedless destruction of our habitat. Macroevolution also reveals that extinction isn't just a bad dream of the distant past; it is a gathering danger for our world.

Why should we care about what paleontologist Peter Ward tells us about the world's greatest extinction event? No, it's not the extinction of dinosaurs 65 million years ago. That was impressive enough, with an extinction toll of half of the Earth's diversity. Something far more devastating took place 252 million years ago: the Permian extinction, an event that wiped out over 90 percent of the world's species. Our own ancestors barely squeaked by. Ward has argued that our planet may be a beautiful home, but it's also a deadly dangerous one. The Earth has a record of periodically assassinating great swaths of life in mass extinctions. It didn't take an asteroid collision to end the Permian. The agent of death was the planet itself, through changes in geological processes that lowered oxygen levels and through massive volcanic eruptions that released greenhouse gasses that heated the planet and helped stress environments beyond the breaking point. These are rare natural events, but we humans now have the power to trigger another mass extinction through the heat stress we are placing on the biosphere and by our overexploitation of an already overcrowded environment.

This time, we are at risk of becoming our own assassins. The evidence suggests that a mass extinction of species has begun already, with a great growing ripple of organisms that are already in danger. Plants, amphibians, fish, and large mammals are all under threat, as are millions of species of small, obscure creatures that inhabit special environments like rain forests and coral reefs, and help keep the Earth's ecosystems balanced and functioning. Where documented, declines are horrendous. Something like 20 percent of plant species and mammals, and 40 percent of amphibians, are currently endangered. Even turtles, those ancient, lovely, and harmless creatures, are being pushed over the edge as they are hunted for food, collected for the pet trade, and have their habitats destroyed. A full 40 percent of turtle species are now at risk of extinction. Some will go extinct in the next decade. The East Asian giant softshell turtle is down to its last lonely individual. Turtles are not only appealing creatures, they are evolutionary gems, among the most unusual of animals in body structure. They show how developmental processes can evolve to yield novel body plans, and they have a long evolutionary history. The first turtle gazed upon the first dinosaur. What a tragedy if these beautiful creatures, this spectacular evolutionary experiment, should be driven out of existence through our carelessness.

But it's not just frogs and turtles or trees and tigers. The human assault on life on Earth has accelerated to the point where the pressures of humanity threaten the planet's ability to support us. Research on evolution, living and extinct diversity, and ecology have given us the potential to predict and perhaps avert our own catastrophic destruction of the natural world. Sadly, we seem determined to squeeze out the last dime of profit from nature, no matter what. In the vast oceans, our over-fishing is converting marine systems from fish-dominated to a kind of jellyfish world unseen for half a billion years. Sci-fi monster nonsense? No, it's bizarrely true. As we sweep the seas clean of large fish, swarms of large jellyfish have taken over swaths of ocean. By large jellyfish, I don't mean large like grapefruits or even soccer balls. The jellyfish in these ocean areas can be a meter or two across. Once large fish controlled the proliferation of the jellies, but now the large jellyfish have turned the tables on the flashy vertebrates and eat young fish, keeping fish stocks down. If a jellyfish ocean seems like a step back in time, imagine the effect of fertilizer runoff from rivers. Our excess fertilizer is creating marine low-oxygen dead zones across the planet where anaerobic microbes control the agenda. A return by these zones is a reminder of the state of the Earth two billion years ago. We can cast our ever-greater nets upon the water and leave behind a strange depleted sea. Or we can take steps to regulate fisheries and allow stocks to recover. An encouraging report by Worm and co-workers in 2009 showed that in some parts of the world, steps to reduce exploitation have resulted in recoveries of fish stocks. Less than a quarter of fish stocks are involved. Action must be taken globally.

John Man, in his book *Gobi,* made an observation relevant to our culture: "In the West, urbanization overlies nature, reducing it from being a foundation for life to a mere adjunct." The "West" seems to get ever larger. We are undermining the habitats of some of the world's most important creatures. We are with thoughtless self-centered parochialism making changes of vast magnitude in the planetary systems that allow humans to live on Earth. The commonly held view is that research on medical problems is most critical to the human condition. That focus might disguise a greater threat. Perhaps a shift to include the sciences of natural history is in order. Understanding of the ecosystems, organisms, and extinctions of the past and present become more and more pressingly relevant to our survival and health.

We may see a loss through extinction of perhaps half the animal species of our planet within the next century. As this is the only planet we have at our disposal, it should give us some pause when we think about our slash-and-burn business-as-usual approach. No one should be lulled by notions of having our surplus human population living in giant rotating space stations or moving to domed colonies on Mars. It might be appealing to think of migrating to the open frontier of Mars for a fresh start, but the planet's living conditions resemble the dry, cold valleys of Antarctica – without the pleasure of breathable air. Our world is rich in life. Do we want to live in a world where we interact with only a handful of crop species plus weedy animals like rats, pigeons, roaches, and head lice? Do we want to live on artificially flavored human kibble commercially produced from vats of algae?

How much of nature is it safe to dispose of? We really don't know, but let's remember that only two food levels are needed for ecological stability – photosynthetic producers such as plants and the algae of the sea convert solar energy to the chemical energy of carbohydrates. They feed the decomposers of dead organic material, bacteria, and fungi that live by turning everything back to gas. That's it. We are not necessary for a basic stable planetary biosphere. In fact, that was the condition of life on Earth for the first monotonous 85 percent of the planet's history. For the first three-plus billion years of life on Earth, this was a world of dreary greenish-yellow microbial mats covering the sea floor in immense slimy carpets and humps. A toxic atmosphere lacking oxygen prevailed much of that long era. We have the fossils of that world, the stromatolites. They had it good in those long-off times. Some fossil microbial and algal mounds are as big as houses.

Animals are really only a phenomenon of the last 550 million years out of the Earth's 4.6-billion-year history. Life originated about 4 billion years ago, and our own species has been around for a mere 200,000 years. If you wanted to celebrate our late, short tenure with some red-blooded vigor, you could climb a twenty-thousand-foot mountain starting from sea level. If that mountain's height represents the history of life on Earth, starting at its base, you would have to haul yourself 19,999 feet up the mountain, to the last exhausting foot, in order to see the nubbin of rock that represents the tiny fraction of the time since life originated that we have been here as a species. Humans and the other animals just provide a decorative frill on the mural of life. We are mobile bric-a-brac dancing over the surface of the hidden real masters of the planet – microbes. Our microbial friends are happy to include us in the menu, but we're really not the main course in the cycling of material and energy in their world. Global changes now include the increase of anaerobic microbe-dominated parts of the oceans – huge oxygen-poor zones deadly for animals. Observations like these are inconvenient to our self-esteem, and the response of narrow interests is to suppress the inconvenient.

Lack of attention to what already has been observed isn't limited to biology. There is more that we just don't want to know. The Swedish chemist Svante Arrhenius discovered the fundamental principles of how CO_2 works as a greenhouse gas a century ago. The writings of the leading climate scientist, James Hansen, make clear that the science is well understood. Hansen's approach of using a kind of paleontology, which reveals the history of atmospheric gas composition preserved in Arctic and Antarctic ice sheets, shows that the pattern of temperature rises and falls over the past 800,000 years is closely linked to CO_2 levels. Increasing CO_2 in the atmosphere beyond what can be absorbed by the oceans will raise the Earth's temperature. The temperature rise is happening. It is easily measured, but denial is rising in the face of the evidence. A July 2009 article in the British newspaper *Observer* revealed that the Bush administration had classified as secret satellite images showing the fast pace of melting Arctic sea ice. For example, pictures taken of Barrow, Alaska, in 2006 show offshore ice. The ice is absent in a picture taken of the same area in July 2007. This was not just a local or freak event or a bias on the part of the photographer. More than a million square kilometers of sea ice overall had vanished in 2007 compared to what was present in 2006. Summer ice levels for 2008 showed no recovery. Hiding the images is "creating your own reality," a term of hubris coined with a straight face by a Bush administration official. A happy outcome is not likely if nature refuses to operate according to our preferences.

Science is ever more necessary for any solution to our dilemma, the coming global warming or, perhaps more accurately, global heating crisis. The rate of CO_2 accumulation has increased inexorably over the last fifty years to a level not experienced by the Earth in hundreds of thousands of years. Each decade since 1980 has seen a rise in global temperature. The past decade is the hottest in the tally. The year 2010 tied for the hottest year ever recorded. Still, the political will to face up to the crisis is lacking, and we while away time that can never be regained. The term "global warming" is easily robbed of its punch by deniers using fantasies of a gentle warming to suggest it merely means nicer winters and desirable northern destinations for beach holidays. Picture a romantic Arctic Ocean summer resort where you can relax on the beach some endless sunny summer day. Oil rigs, tanker oil spills, and mines leaching toxic metals are the more likely future attractions for Arctic coasts. In reality, the effects of global warming will degrade the habitability of much of the world currently home to people.

The Arctic is, ironically, the fastest warming part of the planet. That is where the ever more obvious early effects of global warming taking place are already having a severe effect on people and nature. As the Arctic permafrost melts, it is releasing stores of methane accumulated over thousands of years. This greenhouse gas is twenty-five times more potent than CO_2. Ice reflects the Sun's heat. Open water welcomes it. Sea ice is vanishing. The Inuit cannot hunt. Even the polar bear swims for its life. We have released a dangerous giant, a positive feedback process that can carry global heating beyond what our release of CO_2 can do alone. The crisis will be a tough test of whether society can mobilize scientific information to enable our survival. The current indicators are that we aren't really paying attention yet.

Climate scientists continue to record the rise in global temperature and are sounding a warning that governments can no longer ignore the climate science in their political and economic decisions. Not with much effect, as this is what former vice president Al Gore called an "inconvenient truth." Another equally inconvenient reality is that the world has too large a human population and is too intertwined and dependent on technology to just dump science. A simpler idyllic life where we all live as foragers in harmony with nature is not going to happen. It will not be possible for millions of people to just shoulder their shotguns and deer rifles and stride off into the woods and hills to independently live off the land when the global trade in food fails. Any such hunter-gatherer self-sufficiency is a fiction, a silly fantasy that is impossible with population size that exceeds the planet's ability to supply our present consumption for very long.

Despite the crucial role of limiting population to build a sustainable human biosphere, the topic of population seems to be a great political taboo. Serious public discussions of the challenges of the threat of global climate change and energy policy somehow shy away from mentioning the effects of uncontrolled growth in human numbers. Yet this issue lies at the real heart of sustainability. The assumption seems to be that we can make our use of the world's resources more efficient and we can develop sustainable power sources. These are goals that must be achieved. But they are not enough if the population of the world continues to grow as it is doing now. The world's human population increased two and a half times between 1950 and 2008. As far as I know, the planet did not increase in size, or in wealth of resources, during that time interval. Nor will it in the next few years as we add the next billion people. No technological progress can keep ahead of the resulting increase in demand, even assuming that global warming does not drastically degrade the world's ability to feed itself – not a strong or safe assumption. We will have to choose to inhabit a world with huge populations of desperately poor people or to reduce the world's population to a level that allows all of us to live well but in a sustainable way.

Mainstream economists and politicians have seen growth as the engine of prosperity. They have foolishly cheered on more growth, but we have likely reached and surpassed the limits of growth. A new kind of economics that creates prosperity is necessary. I don't think that immediacy of that need has gotten through to many people so far. It may not seem that way from our locally favored lives as Americans, but we are presently mining fossil energy and much else that is not unlimited and not being renewed. A few ecological economists such as Herman Daly discuss limiting population and argue for the need for sustainable use of natural resources. Daly points out the simple principles: the biosphere is finite, resources are limited, and the world cannot absorb infinite waste. The discussion is growing. Economist Peter Victor in his 2010 article "Questioning Economic Growth" points out that over the twentieth century, human consumption increased 800 percent. He notes that the politically favored policy of economic growth constrains humanity's progress. The economic discussion in itself is not enough. The whispered conversation is the one about human population size and growth. Humans can only survive in the long run if birth rates equal death rates and if population size stabilizes at a level that allows resource use to equal what can be replaced or sustained. Once we have caught all the fish, building a bigger fishing fleet to chase down the last survivors will not increase the catch. To emphasize the point, in January 2011 a giant bluefin tuna sold for a record $390,000 at the Tokyo fish market. Demand is not abating, and Atlantic tunas will be functionally extinct before 2015. But the conventional acceptance of the magical benefits of growth and more growth in population and consumption still dominates political thinking and the basis for how economies are managed.

DENIAL AND DELUSION

Conservative religion attacks public comprehension of science through its rejection of evidence-based thinking, naturalism, and evolutionary biology, but it's not the only offender. Politicians and commercial interests have become highly effective as saboteurs of people's understanding of science. Politicians are far more likely to discuss religion than science, signaling what they think is important and displacing rational discussion. No wonder. Science denial is doubly profitable, simultaneously scoring points with the religious Right and scooping up campaign "contributions" from corporations invested in producing CO_2. As for corporate interests, Naomi Oreskes and Erik Conway, in their eye-opening book *Merchants of Doubt,* discuss the honing of antiscience expertise starting with the fabrication in the 1950s of the controversy as to whether smoking causes cancer. The antiscience innovation was the use of industry-funded "experts" who muddied the discussion of smoking as causing cancer by claiming that the science was not yet established, that there were other scientific data, and that more time was needed for research. The denialists were well paid by industries with wallets fat enough to buy all the "experts" they cared to deploy. Tobacco companies provided the cash in the smoking controversies, and oil companies pay well for global warming denial. The motive for tobacco companies was a cynical desire for profits despite the fact that their product killed their customers through lung cancer. Of the manly, tough Marlboro Man actors of the cigarette ads of my youth, three have died of lung cancer. Corporate and conservative opposition to antismoking legislation led to the creation of a form of science denial wherein a small group of politically motivated scientists would declare that the science was incomplete and inconclusive. Of course the corporations, in this and other denialist activities, have over the decades wanted to keep their involvement out of the limelight, so well-funded front groups were created with elevating names such as the Committee for a Constructive Tomorrow, the George C. Marshall Institute, the Heartland Institute, and the Advancement of Sound Science Coalition.

The theater of a rapid-fire face-off of our expert versus theirs makes good television, but it only inflates the importance of fringe views and generates confusion. The confusion is intended. This approach delayed tobacco legislation for decades, when the science was already clear by the late 1950s. The same approach was taken to prevent a ban on ozone-destroying chlorofluorocarbons (remember Freon?) and is now again being used against action on global warming. The result transcends any particular example, as the outcome is a degradation of the public understanding of scientific issues and a loss of respect for the scientific process. An extreme mentality that values untrammeled "free enterprise" over human and environmental good seems to urge climate change deniers to battle the specter of Karl Marx while religious deniers of evolution fight the good fight against Darwin, and the twain somehow meet. Adherents of the same religious Right groups who oppose evolution also now rail against acceptance of the well-documented science that shows the heating of our planet driven by the burning of fossil fuels. Christian fundamentalism aligned with free-market fundamentalism – a match made in heaven. It's hardly a surprise. Commentator David Frum made the observation that "American populism has almost always concentrated its anger against the educated rather than the wealthy."

Rationalism will inevitably fly in the face of self-interest and denial because realizing that we have to solve the problems we have created will mean that unpalatable changes will be required of human economic behavior and of attitudes on population size limits to create a sustainable world. The reaction by deniers of climate change and of the entire concept of an environmental crisis is to denigrate the science. Thus one prominent conservative radio broadcaster in 2009 gave the world this piece of sledgehammer profundity: "So we have now the Four Corners of Deceit, and the two universes in which we live. The Universe of Lies, the Universe of Reality, and The Four Corners of Deceit: Government, academia, science, and media." This incoherent and divisive diatribe is vacuous blather, but dangerously nihilistic ideas also emerge from the mouths of supposedly responsible politicians. Representative John Shimkus of Wisconsin thinks that reduction of CO_2 emissions would be harmful because carbon dioxide "is plant food." What source of experience and structure for democratic solutions remain for listeners exposed to all this uninformed bumpf? Being guided by charlatans, radio talking heads like the above speaker, and ubiquitous TV evangelists doesn't just rob people of facts, it abolishes rational thought. The effects are not trifling. Thus, by 2011, Tea Party Republicans in the Congress had drunk so deeply and long from the water trough of antiscience ignorance that their opposition to funding of efforts to control global climate change had frozen into an unalterable implacability.

Denialists have two immediate and ominous impacts. They dumb down the public understanding of science, and they have a disastrous effect on policy. Policy decisions come from politics. People rarely act politically simply based on what scientists write in research papers and reports. Thus communication between scientists and the public are critical. Science has to be made clear to nonscientists, and so do public consequences and possible solutions. A few scientists, such as Rachel Carson, whose 1962 book *Silent Spring* had an immense influence on policies controlling pesticides in the environment, have done that brilliantly. But communication is difficult under the best of conditions, and it can become nearly impossible in the face of counter-arguments and "facts" put out by self-serving expert deniers working on behalf of masked special interests. Denialists are making the United States lose its ability to innovate, so that, for example, when we might be able to create new infrastructure and deploy profitable new technologies to ameliorate global warming we fudge and watch China take the lead. Ignorance and purposeful confusion about concepts and facts are the best political allies science deniers have. Unfortunately, although we might be able to negotiate with people, we can't negotiate with the laws of physics, chemistry, and biology.

HOPE

Stories about the Scopes Trial and the long shifting battle with creationism might suggest that Americans don't like science. That's not the case. There is a great liking for utilitarian science, particularly when it's linked to medicine. People also find a high entertainment value and fascination with basic science as shown by a steady stream of television programs on nature, dinosaurs, and space. Science has been well funded by Congress, and although much of the focus is on medical science, a good deal of funding has gone to basic science, which includes such completely nonutilitarian projects as seeking extrasolar planets and life on Mars. Basic science funding even includes evolution. The ambiguities enter when science contradicts other interests. Where the findings of science support popular views, science has high prestige. When scientific discoveries run counter to favored social or religious views, financial interests, or political advantage, there is a deliberate fomenting of negative reactions, and we discover that antiscience and unreason have plenty of good friends.

The classroom teaching of basic sciences thus is more important than ever. The growing success of various flavors of creationism is only one sign that helping the public understand the nature of science is a vital and a never-ending challenge, and one that is not automatically won. Think of what miseducation can do. Here's an example. The Christian nationalist high school textbook *America's Providential History* tells students, "The secular or socialist has a limited resource mentality and views the world as a pie (there is only so much) that needs to be cut up so that everyone can get a piece. In contrast, the Christian knows that the potential in God is unlimited and that there is no shortage of resources in God's earth. The resources are waiting to be tapped." This dangerous illusion isn't limited to one egregious book. *New York Times* environmental writer John Broder reports the recent words of one slightly muddled denier at a congressional candidate forum who rejected the idea that burning oil can affect global climate: "I cannot help but believe the Lord placed a lot of minerals in our country and it's not there to destroy us."

These delusions now influence legislation – or lack of it.

A 2009 poll of Americans reported that 55 percent believe in angels, whereas only 36 percent accept evolution and 39 percent accept that humans are involved in global warming. There are no measurable data to confirm the existence of angels, but they are more strongly believed in than are well-supported research discoveries. This gives pause to any hope that we are close to being able to stop the slide of the planet into dangerous, perhaps irreversible, change. The motives for the attacks on evolution we know, but why the denial of global warming? I can only presume that the severity of demands to change frightens people, and the disinformation campaigns of corporations interested in profiting from extracting the last ounce of buried carbon from the ground are more powerful than mere data. Even so, the elephant is in the room.

Along with creationism, other movements, such as global warming denial, alternative medicine, and the anti-vaccination campaign, thrive on ignorance of science and have well-documented and unpleasant affects. Even the most curiosity-driven university scientists have another duty beyond discovery. We have to explain to our classes the meaning of science in deciding the fate of our world. Understanding evolution is not so ethereal when it comes to the mushrooming problem of antibiotic resistance or the potential for mass extinction as we give the final careless push and watch the plunge to oblivion of charismatic creatures and the destruction of the very environments that support us as living organisms. We may be among the last to see polar bears, tigers, pandas, elephants, and thousands more species large and small. Some suffer because desperate people are under pressure to survive. Our closest relatives, chimps, gorillas, and orangutans are in danger because of extensive hunting of these creatures as "bush meat" by people on subsidence diets and hungry for protein. Some are heedlessly destroyed, like migratory song birds killed by the thousands in Mediterranean countries, for little more than sport and snacks. Many other species, including primates, are being brutally wasted by uncontrolled habitat destruction for short-term gain.

Some writers have starting to refer to our times as what atmospheric scientist Paul Crutzen baptized as the "Anthropocene," the age when individually puny humans acting in concert with technology have become a powerful geological force. Our actions will determine what happens to the air, seas, and land. We will decide the fates of a vast number of other species. Apparently, as a result of our activities, even the marine algae that produce so much of the oxygen we breathe are declining drastically. We are heating up the planet as we enter on a casually offhand and unplanned trip into the unknown. Not the entirely unknown, though. A picture of past mass extinctions having been triggered by runaway global warming events is emerging from the geological and fossil record. The realization that the mechanisms and resulting events of past mass extinctions apply to modern world should be sobering. Global warming deniers cite the financial costs of giving up a coal- and oil-burning energy system, but King Canute's command could not stem the rising tide and, regrettably, economics can't revamp physics for our convenience.

It is not a good gamble to bet on a heavenly safety net if we get things wrong and have to face the natural consequences. Just ask the residents of Pompeii, now so graphically preserved as hollow human-shaped molds in hardened volcanic ash, or the hundreds of thousands of people who drowned in the Indonesian tsunami a few years ago. It is not a good idea to fly with a pilot who says a prayer instead of reading the weather forecast, looking over his plane, and attending to his pre-flight checklist. If the previous paragraphs seem less than optimistic, it is because of our actions that a vast number of other species face enormous challenges – life and death challenges. We face them too. We are a part of nature, and our power to destroy the rest of nature doesn't give us power over natural laws, nor necessarily give us the ability to survive our mistakes.

I have had the greatest fortune of actually being paid for doing what I love best, associating with insects, sea urchins, fossils, bacteria, genes, and other scientists, asking questions and trying to explain how the exuberant forms of life in our world have come into being. I have been allowed the extraordinary freedom to do the research that I find intellectually interesting and important, and I have been free to teach what I consider crucial for students to understand about biology and our place on this unique planet. In return, I think the best contribution I am able to make in the face of what is a developing evolutionary crisis is in my evolution course for undergraduates. Teachers of biology each have the potential to stand before a few thousand students during our careers. Most of my students will not go to graduate school to study evolution or ecology – perhaps a few will, and I cherish them. Others will become physicians or enter other professions not necessarily science related. It is crucial that they and the other nonscientists of their generation understand that the link between evolutionary biology and human survival is real and consequential. My first goal is to teach my students the beauty of life, the intricate mechanisms by which it works, and its spectacular evolutionary history.

It's also vital to present a critical philosophical foundation as well, to try to be sure that students understand what science is and the distinction between science and religion. It's essential that we do this given the noisy prevalence of evolution denial in our society. I address this issue at the beginning of the course. I let them know why religion has no place in science and why the natural world has to be understood solely as the product of natural laws. That separation includes more than the emphatic denial of creationism in its various guises. I also let them know that science cannot prove or disprove God. There is no point in forcing students to think that they have to take sides between science and personal religion. Indeed, it is not my right to directly question students' religious beliefs or impose on them my own personal religious views. But that's a simpleminded hope that can only work if students don't belong to a fundamentalist sect that proclaims the literal truth for its creation stories. I don't tell them the Bible is a book of fables, but from the point of view of fundamentalist students who hold creationism as an integral part of their faith, I'm doing just that. By discussing science and upholding its integrity, I inevitably tap into conflicting worldviews. Without maintaining naturalism in science, any sensible discussion of the natural world would be impossible. I hope some of my creationist students do more than just learn the material and give it back to me on exams. The thinking is for them to do.

The philosophical foundations of science include issues beyond the distinction of science from nonscience. It is vital for students to learn that science is a process, a process of asking questions that can be addressed step by step. Most early education in science is about learning facts. The problem is that facts are like mummies, old and dry. Science is alive and moving. I pose my lectures as questions, such as "How do we know the Earth is old?" That changes the framework from fact to process and lets me explore how we come to know something so basic. It turns out that very few students come into the class with the background that would allow them to give the answer, but it's vital to accepting and understanding evolution. I teach the basic science that gave us the age of the Earth, the evidence for the evolution of organisms, the existence of transitional forms in the fossil record, and the evidence of the working of natural selection. Finally, it has become increasingly important to let students know that evolutionary processes aren't limited to fruit flies or remote island finches, but that all living organisms including ourselves are the products of evolution and that evolution has a continuing impact on human life. We can see, if we have open eyes and minds, how the rapid evolution of disease organisms and evolution of antibiotic resistance puts us at hazard. I discuss as well what the past shows us about the potential of massive human impact. For good or ill the Anthropocene is here. We have now become the dominant force in crucial planetary processes, and we have to discuss that as part of a course in evolution.

Darwin left an uplifting message as the last paragraph in *Origin of Species,* his famous image of a tangled bank made up of the grandeur of life produced by natural selection. A lot has changed since 1859. But even under the human tread, there are still many tangled banks. Some lie near our communities in still-undisturbed and protected local places. They comfort us as much as they intrigue us. Others cling on in the threatened great treasures of life on Earth, the vast Amazon rain forest, the Indonesian rain-forest homes of the last orangutans, the African forests and savannas, the mountains of West Virginia, and the Great Barrier Reef of Australia (not all tangled banks lie on land). The tangled bank still lives, but it is now up to us to make sure that it is not devastated by a thousand local wounds, covered by asphalt parking lots and banal shopping malls, dynamited for coal, illegally logged, bulldozed, developed, poached, drilled, or polluted by industry or giant oil "spills." Nor should we allow the big threats that have the potential to eclipse living things on a global scale. One entire ecosystem, coral reefs, a glorious outcome of tens of millions of years of evolution, is in danger of disappearing by the end of the century. Not even local tangled banks can survive independently of global events. It is our responsibility not to allow the planet to undergo a degree of global warming that threatens all of nature – us included. Now humans have become a force that rivals all the causes of past mass extinctions. Our new role as a geological agent affecting the Earth's surface, oceans, and atmosphere is not an abstraction or a plot device in a science fiction movie. Our actions threaten vast domains that we thought we could never control or destroy – the Arctic, the tropics, the seas, and the air itself.

Despite all this, I want to say that I have great enthusiasm for life and hope for the future. We have the capacity, and still the time, if we start on controlling the attitudes and agents we have unleashed in taking an extractive approach to nature. With public and political will, a sustainable human culture can be achieved. Humans can live in concert with the tangled web and can live vastly more comfortably than most people on Earth do now.

The time for action is, however, growing short. Population size is better addressed now than later, because people are not just numbers. They have to be accommodated once born, and each of us leaves a footprint. In addition, extinctions pile up and greenhouse gasses lead to threshold effects that can't be easily reversed. Again, action now is much more feasible now than once irreversible changes have begun. Biology, from bacteria dividing merrily in a finite culture tube, cattle grazing on the commons, or past societies undergoing ecological collapse, shows that non-sustainable growth leads to collapse. We don't have the choice not to limit population size and planetary exploitation. We only have the choice of how to do it, by allowing catastrophic collapse or by making workable choices. A sustainable world means taking reality seriously and going beyond the tentative and feeble steps taken so far. It is our responsibility, if we hope to live long and well as a species, to curb our exploitive use of the planet and to control our population size such as to leave our descendants the possibility of both prosperity and sustainability. I hope that we have the wisdom to cherish and keep alive the glory of life that evolution has given us.

Selected Bibliography

PREFACE

Crump, M. 2000. *In Search of the Golden Frog.* Chicago: University of Chicago Press.

Darwin, C. 2004. *The Voyage of the Beagle.* Washington, D.C.: National Geographic Society.

Leakey, L. S. B. 1966. *White African: An Early Autobiography.* Cambridge, Mass.: Schenkman.

Lowman, M. D. 1999. *Life in the Treetops: Adventures of a Woman in Field Biology.* New Haven: Yale University Press.

Vermeij, G. 1997. *Privileged Hands: A Scientific Life.* New York: W. H. Freeman.

CHAPTER 1. SPACE-TIME

Fortey, R. 2004. *Earth: An Intimate History.* New York: Alfred A. Knopf.

Fraser, E. 2008. The World War II "Thomas Cook" undercover mail service between Canada and Norway. *Posthorn,* May, 3–10.

Mark, H., and R. Raff. 1941. *High Polymeric Reactions.* New York: Interscience Publishers.

Wacey, D., M. R. Kilburn, M. Saunders, J. Cliff, and M. D. Brasier. 2011. Microfossils of sulphur-metabolizing cells in 3.4-billion-year-old rocks of Western Australia. *Nature Geoscience.* 4:698–702.

Zalasiewicz, J. 2010. *The Planet in a Pebble: A Journey into Earth's Deep History.* Oxford: Oxford University Press.

CHAPTER 2. LAYERS OF THE PAST

Arcand, G. 2005. *Biographie du Colonel Gérard Dufresne.* Shawinigan, Quebec: Société d'histoire mauricienne.

Haak, W., O. Balanovsky, J. J. Sanchez, S. Koshe, V. Zaporozhchenko, C. J. Adler, C. S. I. Der Sarkissian, G. Brandt, C. Schwarz, N. Nicklisch, V. Dresely, B. Fritsch, E. Balanovska, R. Villems, H. Meller, K. W. Alt, A. Cooper, and the Genographic Consortium. 2010. Ancient DNA from European early neolithic farmers reveals their Near Eastern affinities. *PLoS Biol.* 8(11):e1000536. DOI:10.1371/journal.pbio.1000536

La Rochelle, F. 1988. *Histoires de Shawinigan.* Shawinigan, Quebec: Privately published.

National Geographic. 1996–2011. Genographic Project. https://genographic.nationalgeographic.com/genographic/index.html/.

CHAPTER 3. AN AGE OF DINOSAURS

Brinkman, P. D. 2010. *The Second Jurassic Dinosaur Rush: Museums and Paleontology in America at the Turn of the Twen-*

tieth Century. Chicago: University of Chicago Press.

Chiappe, L. M. 2007. *Glorified Dinosaurs: The Origin and Early Evolution of Birds*. Sydney: University of New South Wales Press.

Fox, R. M., A. W. Lindsey Jr., H. K. Clench, and L. D. Miller. 1965. *The Butterflies of Liberia*. Memoirs Amer. Entom. Soc. 19. Philadelphia.

Klein, N., K. Remes, C. T. Gee, and P. M. Sander, eds. 2011. *Biology of the Sauropod Dinosaurs: Understanding the Life of Giants*. Bloomington: Indiana University Press.

McGinnis, H. J. 1982. *Carnegie's Dinosaurs*. Pittsburgh: Carnegie Institute.

Milner, R. 2012. *Charles R. Knight: The Artist Who Saw Through Time*. New York: Abrams.

CHAPTER 4. A SCHOOL A MINUTE

Alvarez, W. 1997. *T. rex and the Crater of Doom*. Princeton: Princeton University Press.

Levy, D. H. 2000. *Shoemaker by Levy. The Man Who Made an Impact*. Princeton: Princeton University Press.

Powell, J. L. 1998. *Night Comes to the Cretaceous: Dinosaur Extinction and the Transformation of Modern Geology*. New York: W. H. Freeman.

CHAPTER 5. IN THE NATURAL WORLD

Al-Khalili, J. 2011. *The House of Wisdom: How Arabic Science Saved Ancient Knowledge and Gave Us the Renaissance*. New York: Penguin Press.

Conniff, R. 2011. *The Species Seekers: Heroes, Fools, and the Mad Pursuit of Life on Earth*. New York: W. W. Norton & Co.

Darwin, C. 1859. *The Origin of Species*. London: John Murray.

Desmond, A. 1994. *Huxley: The Devil's Disciple*. London: Michael Joseph.

Dray, P. 2005. *Stealing God's Thunder: Benjamin Franklin's Lightning Rod and the Invention of America*. New York: Random House.

Erwin, T. L. 1982. Tropical forests: Their richness in coleoptera and other arthropod species. *Coleopterists Bulletin* 36:74–75.

Freely, J. 2009. *Aladdin's Lamp: How Greek Science Came to Europe Through the Islamic World*. New York: Alfred A. Knopf.

Gillispie, C. C. 1951. *Genesis and Geology*. Cambridge: Harvard University Press.

Gosse, E. 2004. *Father and Son*. Edited by Michael Newton. New York: Oxford University Press.

Hamilton, A. J., Y. Basset, K. K. Benke, P. S. Grimbacher, S. E. Miller, V. Novotny, G. A. Samuelson, N. E. Stork, G. D. Weiblen, and J. L.Yen. 2010. Quantifying uncertainty in estimation of tropical arthropod species richness. *Amer. Nat.* 176:90–95.

Huxley, T. H. 1860. Darwin on the origin of species. *Westminster Reviews*, n.s. 17:541–570.

———. 1860. From a letter to Charles Kingsley on death of Huxley's four year old son. September 23. In *The Life and Letters of Thomas Henry Huxley*, by Leonard Huxley. Project Gutenberg. http://www.gutenberg.org/wiki/Main_Page/.

Ladouceur, R. P. 2008. Ella Thea Smith and the lost history of American high school biology textbooks. *Jour. Hist. of Biol.* 41:435–471.

Moon, T. J., P. B. Mann, and J. H. Otto. 1956. *Modern Biology*. New York: Henry Holt.

Mora, C., D. P. Tittensor, S. Adl, A. G. B. Simpson, and B. Worm. 2011. How many species are there on earth and in the ocean? *PLoS Biol.* 9:e1001127.

Oparin, A. 1953. *The Origin of Life*. Translated by Sergius Mogulis. New York: Dover Books.

Raff, R. A. 1960. Two species of Odonata not previously recorded from Pennsylvania. *Entom. News* 71:262.

Skoog, G. 2005. The coverage of human evolution in high school biology textbooks in the 20th century and in current state science standards. *Science & Education*. 14:395–422.

Stewart, A. 2004. *The Earth Moved: On Remarkable Achievements of Earthworms*. Chapel Hill: Algonquin Books.

CHAPTER 6. TRANSFORMATIONS

deBeer, G. *Embryos and Ancestors*. 3rd ed. 1958. Oxford: Oxford University Press.

Dixon, B. 1994. *Power Unseen: How Microbes Rule the World*. New York: W. H. Freeman.

Gubanov, N. M. 1951. Giant nematoda from the placenta of Cetacea: *Placentonema gigantissima* nov. gen. sp. *Dokl. Akad. Nauk. SSSR* 77:1123–1125.

Thomson, K. S. 1995. *HMS Beagle*. New York: W. W. Norton.

CHAPTER 7. GOING SOUTH

Benjamin, R. 2009. *Searching for Whitopia*. New York: Hyperion.

Collins, F. S. 2004. What we do and don't know about "race," "ethnicity," genetics and health at the dawn of the genome era. *Nature Genet. Suppl.* 36:S13–S15.

Gibbons, A. 2010. Human evolution: Tracing evolution's recent fingerprints. *Science* 329:740–742.

Jorde, L. B., and S. P. Wooding. 2004. Genetic variation, classification and "race." *Nature Genet. Suppl.* 36:S28–S33.

Long, J. C., J. Li, and M. E. Healy. 2009. Human DNA sequences: More variation and less race. *Amer. Jour. Phys. Anthro.* 139:23–34.

Marcus, R. A. 2006. A slide toward segregation. *Washington Post*, November 29.

Tishkoff, S. A., and K. K. Kidd. 2004. Implications of biogeography of human populations for "race" and medicine. *Nature Genet. Suppl.* 36:S21–S27.

CHAPTER 8. LEARNING TO LOVE THE BOMB

Klotz, I. 2009. NASA to start irradiating monkeys. *Discovery News*, October 29. http://news.discovery.com/space/space-radiation-monkeys.html/.

McFadden, R. D. 2010. Samuel T. Cohen, neutron bomb inventor, dies at 89. *New York Times*, December 1.

Wade, N. 1978. India bans monkey export: U.S. may have breached accord. *Science* 199:280–281.

CHAPTER 9. ON THE ROAD TO CHIAPAS

Fry, B. G., N. Vidal, J. A. Norman, F. J. Vonk, H. Scheib, H. S. F. Ramjan, S. Kuruppu, K. Fung, S. B. Hedges, M. K. Richardson, W. C. Hodgson, V. Ignjatovic, R. Summerhayes, and E. Kochva. 2006. Early evolution of the venom system in lizards and snakes. *Nature* 439:584–588.

Haug, G. H., D. Günther, L. C. Peterson, D. M. Sigman, K. A. Hughen, and B. Aeschlimann. 2003. Climate and the collapse of Maya civilization. *Science* 299:1731–1735.

Simpson, G. G. 1953. *The Major Features of Evolution*. New York: Columbia University Press.

CHAPTER 10. THE MASKED MESSENGER

Raff, R. A., H. V. Colot, S. E. Selvig, and P. R. Gross. 1972. Oogenetic origin of messenger RNA for embryonic synthesis of microtubule proteins. *Nature* 235:211–214.

CHAPTER 11: EVOLUTION AS SCIENCE

Lewis, C. 2000. *The Dating Game: One Man's Search for the Age of the Earth.* Cambridge: Cambridge University Press.

Richet, P. 2007. *A Natural History of Time.* Chicago: University of Chicago Press.

Rudwick, M. J. 2005. *Bursting the Limits of Time: The Reconstruction of Geohistory in the Age of Revolution.* Chicago: University of Chicago Press.

Ruse, M. 2003. *Darwin and Design: Does Evolution Have a Purpose?* Cambridge: Harvard University Press.

Shubin, N. 2008. *Your Inner Fish: A Journey into the 3.5-Billion-Year History of the Human Body.* New York: Pantheon Books.

CHAPTER 12. DINING WITH DARWIN

Anonymous chapter top quote. 2010. Student class evaluation comment for a course titled "Heredity, Evolution and Society" and taught by Elizabeth Raff, May.

Apesteguía, S., and H. Zaher. 2006. A Cretaceous terrestrial snake with robust hindlimbs and a sacrum. *Nature* 440:1037–1040.

Cohn, M. J., and C. Tickle. 1999. Developmental basis of limblessness and axial patterning in snakes. *Nature* 399:474–479.

Cook, L. M. 2000. Changing views on melanic moths. *Biol. Jour. Linnean Soc.* 69:431–441.

Friedman, M. 2008. The evolutionary origin of flatfish asymmetry. *Nature* 454:209–212.

Grant, K. T., and G. B. Estes. 2009. *Darwin in Galapagos: Footsteps to a New World.* Princeton: Princeton University Press.

Hanson, T. 2011. *Feathers: The Evolution of a Natural Miracle.* New York: Basic Books.

Jones, S. 2011. *The Darwin Archipelago: The Naturalist's Career Beyond Origin of Species.* New Haven: Yale University Press.

Moyal, A. *Platypus: The Extraordinary Story of How a Curious Creature Baffled the World.* 2001. Washington, D.C.: Smithsonian Institution Press.

Müller, F. 1869. *Facts and Arguments for Darwin.* Translated by W. S. Dallas. London: John Murray.

Nicholas, F. W., and J. M. Nicholas. 2008. *Charles Darwin in Australia.* Cambridge: Cambridge University Press.

Palci, A., and M. W. Caldwell. 2007. Vestigial forelimbs and axial elongation in a 95 million-year-old non-snake squamate. *J. Vert. Paleo.* 27:1–7.

Schoenemann, P. T. 2006. Evolution of the size and functional areas of the human brain. *Annu. Rev. Anthropol.* 35:379–406.

White, T. D., B. Asfaw, Y. Beyene, Y. Haile-Selassie, C. O. Lovejoy, G. Suwa, and G. WoldeGabriel. 2009. *Ardipithecus ramidus* and the paleobiology of early hominids. *Science* 326:75–86.

Zimmer, C. 1998. *At the Water's Edge: Macroevolution and the Transformation of Life.* New York: Free Press.

Zuckerkandl, E., and L. Pauling. 1965. Molecules as documents of evolutionary history. *J. Theoret. Biol.* 8:357–366.

CHAPTER 13. LIFE WITH SEA URCHINS

Abzhanov, A., W. P. Kuo, C. Hartmann, B. R. Grant, P. R. Grant, and C. J. Tabin. 2006. The calmodulin pathway and evolution of elongated beak morphology in Darwin's finches. *Nature* 442:563–567.

Gould, S. J. 1977. *Ontogeny and Phylogeny.* Cambridge: Harvard University Press.

Jeffery, W. R., and R. A. Raff, eds. 1983. *Space, Time, and Pattern in Embryonic Development.* New York: Alan R. Liss.

Raff, R. A., and E. C. Raff, eds. 1987. *Development as an Evolutionary Process.* New York: Alan R. Liss.

Weiner, J. 1994. *Beak of the Finch: A Story of Evolution in Our Time.* New York: Alfred A. Knopf.

CHAPTER 14. EMBRYOS EVOLVING

Aguinaldo, A.M.A., J. M. Turbeville, L. S. Linford, L. Hebshi, M.C. Rivera, J. R. Garey, R. A. Raff, and J. A. Lake. 1997. Evidence from 18S ribosomal DNA for a clade of nematodes, arthropods and other molting animals. *Nature* 387:489–493.

Bonner, J. T., ed. 1982. *Evolution and Development.* Berlin: Springer-Verlag.

Budd, G. E., and M. J. Telford. 2009. The origin and evolution of arthropods. *Nature* 457:812–817.

Carroll, S. B. 2005. *Endless Forms Most Beautiful: The New Science of Evo-Devo and the Making of the Animal Kingdom.* New York: W. W. Norton.

Domazet-Loso, T., and D. Tautz. 2010. A phylogenetically based transcriptome age index mirrors ontogenetic divergence patterns. *Nature* 468:815–818.

Duboule, D. 1994. Temporal colinearity and the phylotypic progression: A basis for the stability of a vertebrate Bauplan and the evolution of morphologies through heterochrony. *Development* (Suppl.):135–1342.

Field, K., G. J. Olsen, D. J. Lane, S. J. Giovannoni, N. R. Pace, M. T. Ghiselin, E. C. Raff, and R. A. Raff. 1988. Phylogeny of the animal kingdom based on 18S rRNA sequence data. *Science* 239:748–753.

Grenier, J. K., T. L. Garber, R. Warren, P. M. Whitington, and S. Carroll. 1997. Evolution of the entire arthropod *Hox* gene set predated the origin and radiation of the onychophoran/arthropod clade. *Curr. Biol.* 7:547–553.

Irie, N., and S. Kuratani. 2011. Comparative transcriptome analysis reveals vetebrate phylotypic period during organogenesis. *Nature Commun.* 2 (March): 248.

Kalinka, A. T., K. M. Varga, D. T. Gerrard, S. Preibisch, D. L. Corcoran, J. Jarrells, U. Ohler, C. M. Bergman, and P. Tomancak. 2010. Gene expression divergence recapitulates the developmental hourglass model. *Nature* 468:811–814.

Raff, R. A. 1996. *The Shape of Life: Genes, Development, and the Evolution of Animal Form.* Chicago: University of Chicago Press.

Raff, R. A., and T. C. Kaufman. 1983. *Embryos, Genes and Evolution: The Developmental-Genetic Basis of Evolutionary Change.* New York: Macmillan.

CHAPTER 15. EVOLUTION IN THE TASMAN SEA

Miskelly, A. 2002. *Sea Urchins of Australia and the Indo-Pacific.* Sydney: Capricornica Publications.

Nilsson, D.-E., and S. Pelger. 1994. A pessimistic estimate of the time required for an eye to evolve. *Proceedings: Biological Sciences* 256:53–58.

Raff, E. C., E. M. Popodi, J. S. Kauffman, B. J. Sly, F. R. Turner, V. B. Morris, and R. A. Raff. 2003. Regulatory punctuated equilibrium and convergence in the evolution of developmental pathways in direct-developing sea urchins. *Evo. Dev.* 5:478–493.

Raff, E. C., E. M. Popodi, B. J. Sly, F. R. Turner, J. T. Villinski, and R. A. Raff. 1999. A novel ontogenetic pathway in hybrid embryos between species with different modes of development. *Development* 126:1937–1945.

Raff, R. A. 2009. Origins of metazoan body plans: The larval revolution. In *Animal Evolution: Genomes, Fossils, and Trees,* edited by M. J. Telford and D. T. J. Littlewood, 43–51. Oxford: Oxford University Press.

Raff, R. A., J. Anstrom, C. J. Huffman, D. S. Leaf, J.-H. Loo, R. M. Showman, and D. E. Wells. 1984. Origin of a gene regulatory mechanism in echinoderm evolution. *Nature* 310:312–314.

Raff, R. A., and M.S. Smith. 2009. Axis formation and the rapid evolutionary transformation of larval form. *Curr. Topics Dev. Biol.* 86:163–190.

Sly, B. J., M. S. Snoke, and R. A. Raff. 2003. Who came first? Origins of bilaterian metazoan larvae. *Int. J. Dev. Biol.* 47:623–632.

Smith, M. S., K. E. Zigler, and R. A. Raff. 2007. Evolution of direct developing larvae: Selection vs. loss. *BioEssays* 29:566–571.

Williams, D. H. C., and D. T. Anderson. 1975. The reproductive system, embryonic development, larval development, and metamorphosis of the sea urchin *Heliocidaris erythrogramma* (Val.) (Echinoidea: Echinometridae). *Aust. J. Zool.* 23:371–403.

CHAPTER 16. AN ALTERNATE PRESENT

Dickman, C. 2007. *A Fragile Balance: The Extraordinary Story of Australian Marsupials.* Illustrations by Rosemary Woodford Ganf. Chicago: University of Chicago Press.

Flannery, T. 2004. *Chasing Kangaroos.* New York: Grove Press.

Tait, N. N., and J. M. Norman. 1999. Novel mating behaviour in *Florelliceps stuchburyae* gen. nov., sp. nov. (Onychophora: Peripatopsidae) from Australia. *J. Zool. Lond.* 253:301–308.

CHAPTER 17. BIOLOGY MEETS FOSSILS

Budd, G. E., and M. J. Telford. 2009. The origin and evolution of arthropods. *Nature* 457:812–817.

Debelle, P. 2003. Fossil discovery a magnet for raiders of the primordial ark. *Theage.com.au,* October 25. http://www.theage.com.au/articles/2003/10/24/1066974325195.html?from=storyrhs/.

Fedonkin, M. A., J. G. Gehling, K. Grey, G. Narbonne, and P. Vickers-Rich. 2007. *The Rise of Animals: Evolution and Diversification of the Kingdom Animalia.* Baltimore: Johns Hopkins University Press.

Raff, E. C., K. L. Schollaert, D. E. Nelson, P. C. J. Donoghue, C.-W. Thomas, F. R. Turner, B. D. Stein, X. Dong, S. Bengtson, T. Huldtgren, M. Stampanoni, Y. Chongyu, and R. A. Raff. 2008. Embryo fossilization is a biological process mediated by microbial biofilms. *Proc. Nat. Acad. Sci. USA* 105:19360–19365.

Raff, E. C., J. A. Villinski, F. R. Turner, P. C. Donahue, and R. A. Raff. 2006. Experimental taphonomy: Feasibility of fossil embryos. *Proc Natl Acad Sci USA* 103:5846–5851.

Raff, R. A. 2007. Written in stone: Fossils, genes and evo-devo. *Nature Rev. Genet.* 8:911–919.

Raff, R. A., and E. C. Raff. 1970. Respiratory mechanisms and the metazoan fossil record. *Nature* 228:1003–1005.

Xiao, S., Y. Zhang, and A. Knoll. 1998. Three-dimensional preservation of algae and animal embryos in a Neoproterozoic phosphorite. *Nature* 391:553–558.

CHAPTER 18. DARWIN'S DAY IN COURT

Bhattacharjee, Y. 2010. NSF board draws flak for dropping evolution from *Indicators. Science* 328:150–151.

Boyer, P. J. 2010. Frat House for Jesus: The entity beyond C Street. *New Yorker,* September 13. http://www.newyorker.com/reporting/2010/09/13/100913fa_fact_boyer#ixzz15xNwanSM/.

Boykin, W., Deputy Undersecretary of Defense for Intelligence. 2003. Reported by the *Los Angeles Times,* October 16.

Bush, G. H. W. 1987. Quotation. Reply by Vice President G. H. W. Bush to *American Atheists* reporter Robert Sherman, Chicago.

Chapman, M. 2007. *Forty Days and Forty Nights: Darwin, Intelligent Design, God, Oxycontin, and Other Oddities on Trial in Pennsylvania.* New York: HarperCollins.

Crumb, R. 2009. *The Book of Genesis Illustrated.* New York: W. W. Norton.

Doolittle, R. F. 2009. Step-by-step evolution of vertebrate blood coagulation. *Cold Spring Harb Symp Quant Biol.* 74:35–40.

Garrison, C. 2009. Evolution t-shirts on trial in Missouri town. *Riverfront Times.com,* Saint Louis News blog, August 31.

Halbrook, S. C. 2001. *God Is Just: A Defense of the Old Testament Civil Laws, Biblical Theocracy, Justice, and Slavery versus Humanistic Theocracy, "Justice," and Slavery.* Theonomy Resources Media. http://www.lulu.com/items/volume_70/10853000/10853587/1/print/god_is_just2.pdf/.

Harvey, S. G., and E. Rothschild. 2010. Defending Darwin. *Litigation.* 37:8–14.

Pallen, M. J., and N. J. Matzke. 2006. From the origin of species to the origin of bacterial flagella. *Nature Rev. Microbiol.* 4:784–790.

Pierce, C. P. 2009. *Idiot America: How Stupidity Became a Virtue in the Land of the Free.* New York: Doubleday.

Scopes, J. T., and J. Presley. 1967. *Center of the Storm: Memoirs of John T. Scopes.* New York: Holt, Rinehart and Winston.

Schaeffer, F. 2011. *Sex, Mom, and God: How the Bible's Strange Take on Sex Led to Crazy Politics – and How I Learned to Love Women (and Jesus) Anyway.* Cambridge: Da Capo Press.

Sharlet, J. 2010. *C Street. The Fundamentalist Threat to American Democracy.* New York: Little Brown.

CHAPTER 19. CREATIONIST MAKEOVERS

Asma, S. T. 2011. Risen apes and fallen angels: The new museology of human origins. *Curator* 54:141–163.

Berkman, M. B., and E. Plutzer. 2011. Defeating creationism in the courtroom, but not in the classroom. *Science* 331:404–405.

Boston, R. 2011. Biological warfare: Battles under way in Texas and Louisiana over science education. *Wall of Separation,* April 29. http://blog.au.org/2011/04/29/biological-warfare-battles-under-way-in-texas-and-louisiana-over-science-education/?utm_source=au-homepage&utm_medium=feed&utm_campaign=Recently-on-homepage/.

———. 2011. From the Kentucky 'Ark Park' to back-door creationist legislation, religious right forces are demanding state support of fundamentalist dogma. Americans United for Separation of Church and State. July 5. http://www.au.org/media/church-and-state/archives/2011/07/creationism-crusade.html/.

Branch, G., E. C. Scott, and J. Rosenau. 2010. Dispatches from the evolution wars: Shifting tactics and expanding battlefields. *Annu. Rev. Genomics Hum. Genet.* 11:317–338.

Harmon, K. 2011. Evolution abroad: Creationism evolves in science classrooms around the globe. *Sci.*

Amer., March 3. http://www.scientificamerican.com/article.cfm?id=evolution-education-abroad/.

Miller, K. R. 1999. *Finding Darwin's God: A Scientist's Search for Common Ground Between God and Evolution.* New York: Cliff Street Books.

Ravitch, F. S. 2011. *Marketing Intelligent Design. Law and the Creationist Agenda.* Cambridge: Cambridge University Press.

Stein, B. 2008. Interview with Paul Crouch Jr., Trinity Broadcasting Network. In science equals murder, by John Derbyshire. *National Review Online,* April 30. http://www.nationalreview.com/corner/162377/science-equals-murder/john-derbyshire/.

Wilson, D. P. 2010. European Christians are at the forefront in accepting evolution: Results from an internet-based survey. *Evo. Dev.* 12:537–650.

CHAPTER 20. EVOLUTION MATTERS

Alley, R. B. 2011. *Earth: The Operators' Manual.* New York: W. W. Norton.

Barnosky, A. D., N. Matzke, S. Tomiya, G. O. U. Wogan, B. Swartz, T. B. Quental, C. Marshall, J. L. McGuire, E. L. Lindsey, K. C. Maguire, B. Mersey, and E. A. Ferrer. 2011. Has the earth's sixth mass extinction already arrived? *Nature* 471:51–57.

Beliles, M. A., and S. K. McDowell. 1989. *America's Providential History.* Charlottesville, Va.: Providence Foundation.

Boyce, D. G., M. R. Lewis, and B. Worm. 2010. Global phytoplankton decline over the past century. *Nature* 466:591–596.

Broder, J. M. 2010. Climate change doubt is Tea Party article of faith. *New York Times,* October 20.

Butchart, S. H. M., M. Walpole, B. Collen, A. van Strien, J. P. W. Scharlemann, R. E. A. Almond, J. E. M. Baillie, B. Bomhard, C. Brown, J. Bruno, K. E. Carpenter, G. M. Carr, J. Chanson, A. M. Chenery, J. Csirke, N. C. Davidson, F. Dentener, M. Foster, A. Galli, James N. Galloway, P. Genovesi, R. D. Gregory, M. Hockings, V. Kapos, J.-F. Lamarque, F. Leverington, J. Loh, M. A. McGeoch, L. McRae, A. Minasyan, M. Hernández Morcillo, T. E. E. Oldfield, D. Pauly, S. Quader, C. Revenga, J. R. Sauer, B. Skolnik, D. Spear, D. Stanwell-Smith, S. N. Stuart, A. Symes, M. Tierney, T. D. Tyrrell, J.-C. Vié, and R. Watson. 2010. Global biodiversity: Indications of recent declines. *Science* 328:1164–1168.

Daly, H. E. 2005. Economics in a full world. *Scientific American,* September, 100–107.

Diamond, J. M. 2005. *Collapse. How Societies Choose to Fail or Succeed.* New York: Viking Press.

Dolan, E. W. 2010. "God won't allow global warming," congressman seeking to head energy committee says. *RAWSTORY,* November 11. http://www.rawstory.com/rs/2010/11/god-global-warming-congressman-energy/.

Ehrlich, G. 2010. *In the Empire of Ice.* Washington, D.C.: National Geographic Society.

Fan, W., T. Yanase, H. Moringa, S. Gondo, T. Okabe, M. Nomura, T. Komatsu, K.-I. Morohashi, T. B. Hayes, R. Takayanagi, and H. Nawata. 2007. Atrazine-indiced aromatase expression is SF-1 dependent: Implications for endocrine disruption in wildlife and reproductive cancers in humans. *Environ. Health Perspec.* 115:720–727.

Flannery, T. 2010. *Here on Earth: A Natural History of the Planet.* New York: Atlantic Monthly Press.

Frum, D. 2010. Post-Tea-Party nation. *New York Times,* November 12.

Gassmann, A. J., J. L. Petzold-Maxwell, R. S. Keweshan, and M. W. Dunbar.

2011. Field-evolved resistance to Bt maize by western corn rootworm. *PLoS One* 6:e22629.

Goldenberg, S., and D. Carrington. 2009. Revealed: The secret evidence of global warming Bush tried to hide. *Observer,* Sunday, July 26.

Grant, J. 2011. *Denying Science: Conspiracy Theories, Media Distortions, and the War Against Reality.* Amherst and New York: Prometheus Books.

Hansen, J. 2009. *Storms of my Grandchildren: The Truth About the Coming Climate Catastrophe and our Last Chance to Save Humanity.* New York: Bloomsbury Press.

Hayes, T. 2004. There is no denying this: Defusing the confusion about atrazine. *Bioscience* 54:1138–1149.

Hayes, T., K. Haston, M. Tsui, A. Hoang, C. Haeffele, and A. Vonk. 2002. Feminization of male frogs in the wild. *Nature* 419:895–896.

Hayes, T. B., V. Khoury, A. Narayan, M. Nazir, A. Park, T. Brown, L. Adame, E. Chan, D. Buchholz, T. Stueve, and S. Gallipeau. 2010. Atrazine induces complete feminization and chemical castration in male African clawed frogs (*Xenopus laevis*). *Proc. Natl. Acad. Sci. USA* 107:4612–4617.

Institute of Science in Society. 2010. Glyphosate resistance in weeds: The transgenic treadmill. *ISIS Report 03/03/10.* http://www.isis.org.uk/glyphosateResistanceTransgenicTreadmil.php/.

Limbaugh, R. 2009. ClimateGate hoax: The universe of lies versus the universe of reality. Transcript. November 24. http://www.rushlimbaugh.com/home/daily/site_112409/content/01125108.guest.html/.

Mann, M. E. 2012. *The Hockey Stick and the Climate Wars: Dispatches From the Front Lines.* New York: Columbia University Press.

Marshall, C. 2011. State legislatures pile onto anti-EPA climate rule effort. *New York Times,* April 1.

McCurry, J. 2011. Huge bluefin tuna fetches record price in Tokyo, but whale is left on the shelf. *Guardian,* January 6.

Oreskes, N., and E. M. Conway. 2010. *Merchants of Doubt: How a Handful of Scientists Obscured the Truth on Issues from Tobacco to Global warming.* New York: Bloomsbury Press.

Otto, S. L. 2011. *Fool Me Twice: Fighting the Assault on Science in America.* New York: Rodale Books.

Pearson, R. 2011. *Driven to Extinction: The Impact of Climate Change on Biodiversity.* New York: Sterling.

Pigliucci, M. 2010. *Nonsense on Stilts: How to Tell Science from Bunk.* Chicago University Press of Chicago.

Powell, J. L. 2011. *The Inquisition of Climate Science.* New York: Columbia University Press.

Richardson, A. J., A. Bakun, G. C. Hays, and M. J. Gibbons. 2009. The jellyfish joyride: Causes, consequences and management responses to a more gelatinous future. *Trends Ecol. Evol.* 24:312–322.

Tollefson, J. 2011. The sceptic meets his match. *Nature* 475:440–441.

Turney, C. 2008. *Ice, Mud, and Blood: Lessons from Climates Past.* London: Macmillan.

Vermeij, G. J. 2010. *The Evolutionary World: How Adaptation Explains Everything from Seashells to Civilization.* New York: St. Martin's Press.

Victor, P. 2010. Questioning economic growth. *Nature* 486:370–371.

Vie, J.-C., C. Hilton-Taylor, and S. N. Stuart. 2009. *Wildlife in a Changing World: An Analysis of the 2008 IUCN Red List of Threatened Species.* Gland, Switzerland: IUCN.

Ward, P. D. 2004. *Gorgon: Paleontology, Obsession, and the Greatest Catastrophe*

in Earth's History. New York: Viking Press.
———. 2007. *Under a Green Sky: Global Warming, the Mass Extinctions of the Past and What They Can Tell Us About Our Future.* New York: HarperCollins.
———. 2009. *The Medea Hypothesis. Is Life on Earth Ultimately Self-Destructive?* Princeton: Princeton University Press.
Winchester, P. D., J. Huskins, and J. Ying. 2009. Agrichemicals in surface water and birth defects in the United States. *Acta. Paediatr.* 98:664–669.
Worm, B., R. Hilborn, J. K. Baum, T. A. Branch, J. S. Collie, C. Costello, M. J. Fogarty, E. A. Fulton, J. A. Hutchings, S. Jennings, O. P. Jensen, H. K. Lotze, P. M. Mace, T. R. McClanahan, C. Minto, S. R. Palumbi, A. M. Parma, D. Ricard, A. A. Rosenberg, R. Watson, and D. Zeller. 2009. Rebuilding global fisheries. *Science* 325:578–85.

Index

RUDOLF A. RAFF is Distinguished Professor and James H. Rudy Professor of Biology at Indiana University and one of the founders of the field of study known as evolutionary developmental biology. He is director of Indiana Molecular Biology Institute, editor-in-chief of *Evolution & Development,* Guggenheim Fellow, Fellow of the American Academy of Arts and Sciences, and Fellow of the American Association for the Advancement of Science. He is author *The Shape of Life: Genes, Development and the Evolution of Animal Form* and author (with Thom Kaufman) of *Embryos, Genes, and Evolution.*

This book was designed by Jamison Cockerham and set in type by Tony Brewer at Indiana University Press and printed by Sheridan Books, Inc.

The text type is Arno, designed by Robert Slimbach, and the display face is Futura, designed by Paul Renner, both issued by Adobe Systems, Inc.